George Hale Barrus

Boiler Tests

Embracing the Results of One Hundred and Thirty-Seven Evaporative Tests, Made on Seventy-One Boilers...

George Hale Barrus

Boiler Tests
Embracing the Results of One Hundred and Thirty-Seven Evaporative Tests, Made on Seventy-One Boilers...

ISBN/EAN: 9783337279578

Printed in Europe, USA, Canada, Australia, Japan

Cover: Foto ©berggeist007 / pixelio.de

More available books at **www.hansebooks.com**

BOILER TESTS;

EMBRACING THE RESULTS OF ONE HUNDRED AND
THIRTY-SEVEN EVAPORATIVE TESTS, MADE
ON SEVENTY-ONE BOILERS, CON-
DUCTED BY THE AUTHOR.

BY

GEO. H. BARRUS, S. B.,

MEMBER OF AMERICAN SOCIETY OF MECHANICAL ENGINEERS,
AND BOSTON SOCIETY OF CIVIL ENGINEERS.

BOSTON:
PUBLISHED BY THE AUTHOR.
GOWING & CO., AGENTS,
70 KILBY STREET.
1895.

PREFACE.

This book relates to a large number of evaporative tests, which were conducted personally by the author. The substance of Parts I and II, which form the body of the work, was originally prepared at the request of the New England Cotton Manufacturers' Association, and embodied in a paper, which was read to that Society. A few copies of the paper were reprinted from the Transactions of the Association, and furnished to parties interested in the subject. The supply of these was quickly exhausted. The favor with which the paper was received, not only by members of the Association, but also by those who have been supplied with the reprints, led to its publication in the present form. In the revised work the arrangement and numbering of the various boilers have been changed, and cuts have been introduced to show the general features of the boilers tested. An index has also been added. These changes supply what was lacking in the paper as first prepared, and give to the work that degree of completeness which is desired in a book of reference.

The author has made, in later tests, investigations of the heating power of various fuels, determining the heat of combustion, for this purpose, by the use of a coal calorimeter. It has been thought desirable, in view of the bearing which these investigations have upon the general subject of boiler tests and the combustion of fuel, to give a brief account of the work in connection with the tests under consideration, and this has been done in the Appendix. There is also presented in the Appendix a portion of the author's paper on "A Universal Steam Calorimeter," which is added partly on account of its bearing on the subject, and partly because it has, in the author's later tests, taken the place of the Superheating Calorimeter, referred to in Part I.

The book is offered to the public in the belief that it is of considerable interest and value to steam engineers, manufacturers, and any persons who are concerned, either directly or indirectly in the burning of coal and the economical generation of steam.

GEO. H. BARRUS.

95 MILK ST., BOSTON,
    May, 1891.

CONTENTS.

PART I.

Introduction,	9
Methods Employed in Conducting the Tests,	13
Explanation of the Tables of Part Second,	19
Discussion of Results,	21
Comparison of Boilers which produce Saturated Steam with those producing Superheated Steam,	22
General Conditions which Secure Economy,	28
Comparison of Different Kinds of Boilers,	36
Comparison of Different Kinds of Fuel,	43
Miscellaneous Discussion,	51
Flue Heaters,	56

PART II.

Tests of Horizontal Tubular Boilers,	67
Tests of Horizontal Double-deck Boilers,	159
Tests of Plain Cylinder Boilers,	174
Tests of Vertical Tubular Boilers,	183
Tests of Cast Iron Sectional Boilers,	211
Tests of Water-tube Boilers,	217
Summary of Tests,	234

APPENDIX.

A Coal Calorimeter,	249
A Universal Steam Calorimeter,	255

PART I.

Boiler Tests.

INTRODUCTION.

In the pursuit of his business as a steam engineer the author has had frequent occasion during the past ten years to make evaporative tests of steam boilers. The interests which have led to the work are of a varied character. Inventors or builders of new types of boilers have been desirous of learning what economical features their improvements possess, and they have called for the results of an evaporative test as a means of comparison with the performance of existing boilers; or, knowing themselves the economy of the improved generator, they wished, for purposes of trade, to corroborate their own testimony of its superiority, by the certificate of a disinterested engineer. Managers of factories have sought advice upon the cause of excessive consumption of coal in their steam plants. To solve this problem it is necessary to ascertain the evaporative efficiency of the boilers, so as to separate the process of generating the steam from that of its use after generation. They have also desired information upon the general economy of their boilers, and upon the kind of fuel which it would be most advantageous to burn, when all the questions of current price, evaporative efficiency and incidental advantages and objections are taken into account — questions which can be determined satisfactorily only by means of an evaporative test. They have had occasion, too, before completing a settlement with the contractor for a new plant of boilers, to call for a test, to ascertain whether in capacity or economy, or both, the boilers do their work in the manner stipulated in the contract. They have also sought information, which only an evaporative test affords, to guide them in the selection or arrangement of a contemplated new boiler plant, or in the renewal of an old plant. Capitalists have called for evaporative tests upon improved forms of boilers, in which they have in view the investment of money, that they may learn from outside investigation whether the inventor's claims are well founded, or whether the improvement is of sufficient merit to warrant the investment. Finally, a large class of persons, bent on seeking information, have had boilers tested, so as to find out whether, in one way or another, the evaporative efficiency could be improved.

In every case, whether the test has been made at the solicitation of the inventor, the mill-manager, the capitalist or the investigator, there has always been one controlling object in view, on the part of the author, and that is, the determination of the true performance of the boiler, and the work has been undertaken in every instance with the unbiased feeling of a student, interested only in getting at the facts.

In this work the number of individual tests which have been made has reached a sum between three hundred and fifty and four hundred. Many of them have been made in the New England States. The boilers have been situated in cotton mills, woollen mills, and other textile factories, paper mills, machine shops, lumber mills, rubber works and public buildings. They have embraced plain cylinder boilers, horizontal return tubular boilers, locomotive boilers, vertical tubular boilers, water-tube boilers, sectional boilers and patented boilers of various kinds. They have been set in brick-work, set without brick-work, set according to patented methods and set in the ordinary way. They have been fired with anthracite coal of various kinds and sizes, bituminous coal of various kinds, anthracite screenings, mixtures of anthracite screenings and bituminous coal, petroleum oil and coke.

The paper is devoted to an analysis of many of these tests. Of the number referred to, those which are omitted are either of a private nature or are devoid of general interest. They were all conducted personally by the author, with the exception of one set of tests, which were made by an assistant who worked under the author's direction.

The tests are so varied in object, in conditions, and in type of boiler, that it has proved advisable to take up each boiler and the tests made upon it, separately, and give, as briefly as a clear presentation of the subject warrants, the general features and dimensions of the boiler, the data and results of the tests, and memoranda of the main points of information which the tests reveal. Given in this form, the reader will find means for studying the subject, if desired, from his own standpoint. The special object of the paper is an examination of the tests as a whole, and a classification of the data and results, so as to put in connected shape the information which they give as to the conditions which govern the economical generation of steam and the best practice in boiler engineering.

The paper is divided into two parts; Part First embracing the analysis of the tests as a whole, and Part Second treating of the various boilers, and the tests made upon them individually.

Before entering upon the proposed examination, the general subject of boiler tests and the methods which were employed in conducting the tests under consideration demand attention.

The object of all boiler tests, first sought, is the determination of the weight of water evaporated by a pound of coal. The object ultimately sought varies in different cases. Some tests have in view the determination of the efficiency of the boiler as a generator of steam, when operated under such conditions of quality of fuel, kind of firing, capacity and general management as will give the best result. These determine the true efficiency of the generator. Others have in view the general result produced by the boiler, however much it is affected by disturbing causes, and they determine the economy with which the fuel is burned, whether the type of boiler, kind of fuel, capacity, mode of firing and management are favorable or unfavorable.

Owing to the two general objects named, boiler tests might be divided to advantage into two classes. The first might be termed the *engineering test*, being of greatest interest to the engineer, who must know the true efficiency of the boiler to intelligently lay out his plans. The second might be termed the *commercial test*, being of most interest to the manufacturer or steam user, who must know how the coal is used which he has to purchase. In conducting the engineering test the boiler should be worked with a standard quality of fuel. It should be clean on both the interior and the exterior surfaces. It should be fired by a skillful fireman, who should use every exertion to obtain the highest efficiency from the fuel. It should be worked at such a capacity as will secure the most favorable result. The boiler should be worked long enough in preparation for the engineering test to heat itself and the material in which it is encased to its normal working degree. The fire should then be burned down to a low point and thoroughly cleaned, a thickness of 2 or 3 inches of clean burning fuel being retained for a foundation upon which the fire of the test is to be started. With this thin fire, the thickness of which is estimated as a basis for commencement, the test is begun. The boiler is then operated in the manner determined upon, for at least 24 hours, and the quality of coal burned and water evaporated is measured. At the close of the test the fire is again burned down and cleaned, and this is done in such a manner that the condition and thickness of the bed of coal on the grate, as nearly as can be estimated, is the same as it was at the commencement of the test. For the commercial test the boiler should be operated in the manner customarily followed, and the measurement of coal and water taken a sufficient number of days to give an average indication of the performance. In factory work a test of one week's duration is of suitable length. No change should be made in the daily hours of running, the mode of firing, the manner of treating the fires at night, nor in the general operation of the plant, the object being to include all these questions in the determination of the main point at issue.

If all tests could be made in accordance with these plans nothing would be left to be desired. The relative efficiency of different generators would be known at once. The manufacturer would have no complaint to make because the results of the test do not correspond to those obtained under working conditions. Unfortunately, no such uniformity of method can be established, on account of the practicable objections and expense involved in carrying on tests of this kind. The 24-hour continuous engineering test is usually out of the question in a mill which uses steam only 10 or 12 hours a day. The six-days' commercial test is not to be thought of by the manager of a mill, who counts the expense for experts' services. The ordinary calls for evaporative tests are for those which can be made in one or two working days, under conditions as near as possible to the existing ones in the locality where the boiler is placed. These must answer, in ordinary practice, for the purposes of both the engineering and the commercial tests in the scheme outlined above.

METHODS OF CONDUCTING THE TESTS.

Nearly all of the tests referred to in this paper are of the last named class. They were made, as a rule, in factories working 10 or 12 hours per day, where the fires are banked at night, and where an interval of about an hour occurs at noon time, when the production of steam nearly ceases. The boilers were generally cleaned, as far as the outside fire surfaces were concerned, by the use of the tube scraper or steam jet cleaner, preparatory to each test. The inside surfaces were not cleaned and generally were not examined. The mode of firing was not always the best, but it was always good; that is to say, the fires were, as a rule, free from " holes " or spots where the live coal had died out, with the exception of cases where fine grades of anthracite coal, such as chestnut No. 2 and pea, were used, in which there is frequently a tendency to this state of things, and the fires were kept at a thickness of not less than five or six inches. When bituminous coal was burned, it was usually fired on the " spreading " system, and the bed of coal was broken up and kept level when required by the use of the slicing and stoking bar.

The method of conducting most of the tests given in the paper is as follows: Preparatory to the beginning of the test, on the night before its commencement, the fires are burned out at the close of the day's work. The furnace and ash-pit doors and the dampers are closed during the night, and, in case only one or two boilers of a plant are tested, the stop valves are also closed, shutting the steam into the boiler. At some convenient time during the night the furnace and ash-pits are cleaned. At five o'clock in the morning, or at a sufficiently early hour to get up steam for the day's work, the new fire of the test is started with enough pine wood to cover the grates, and coal is fired at once, and the boiler brought to normal action. Oftentimes the fire is in working condition before the steam can be disposed of, and a pause ensues until the usual work of the mill begins, during which time the fire is covered with new coal, and the draught choked by shutting the damper and opening the fire doors. During the noon hour, when the works are stopped, the fire is cleaned (if needed) and covered with new coal, and the draught checked in like manner. As the hour for stopping draws near, the fires are carefully burned down, so that at the close of the day's work they are nearly out. As soon as they are finally extinguished, the contents of furnaces and ash-pits are removed and the doors and dampers closed for the night, and in the cases noted the stop valves are shut. When the coal is moist, a sample is selected and dried for 24 hours, and the quantity of moisture is determined by a comparison of the wet and dry weights. A sample is taken of the ashes and unburned fuel removed from the furnaces and ash-pits at the close of the test, and sifted through a screen having $\frac{3}{8}$-inch meshes. The proportion of unconsumed coal which fails to pass through the screen is determined and applied to the total quantity, and the whole weight of unconsumed coal thus found is deducted from the weight of coal fired, to determine the net quantity consumed. The value of the wood used for starting the fires is taken to be equivalent in coal to $\frac{4}{10}$ of its weight.

At the time of starting the new wood fire of the test, the position of the water line as shown in the gauge glass of the boiler is carefully noted, and the pressure indicated by the steam gauge as well. At the close of the test, it is intended that a sufficient amount of water shall be supplied, so that when the time arrives on the next morning, corresponding to the time when the test was started, the position of the water line and the indication of the steam gauge shall be the same as at the beginning. The last observation of the test is thus made 24 hours after the time of the first observation, when the condition of the boiler is presumably the same. If the water line occupies a different position, a suitable correction is applied to the weight of water measured, which is determined by computing the quantity from the volume which it occupies, calculated from the known dimensions of the boiler. If there is a difference of pressure, no account is taken of it, unless the difference is large; in which case, the quantity of evaporation is computed which would take place if the stop valve were opened and a sufficient amount of steam let off to reduce the pressure from the higher to the lower point. In conducting a test according to this method, the small quantity of evaporation which takes place during the night, produced by heat stored in the boiler at the end of the day's run, is included in the total evaporation of the test.

During the progress of the test the feed-water is weighed previous to its supply either to the pump or injector which feeds it. The weighing apparatus consists of two tanks, one of which rests on a platform scale, which is supported by staging raised above the level of the other. The water is first drawn into the upper tank. Here it is weighed and then emptied into the lower tank. Thence it passes into the suction pipe of the pump or injector, which is temporarily changed from its usual connection for the purpose.

All pipes connected with the feeding apparatus, except those concerned in the work of the test, are disconnected; and if this cannot be done, the tightness of valves which must be depended upon is first assured. If the blow-off valve leaks, it is plugged.

In cases where an injector is used, and where there is no heater connected with the boiler, the temperature of the water is taken at its entrance to the lower tank. If a heater is used, the temperature is taken at its entrance to the boiler, a thermometer being set in the feed pipe for this purpose. If both injector and heater are used, the temperature is taken at the tank and at the entrance to the heater, as well as at the entrance to the boiler. To avoid breakage, the thermometer is placed in a cup filled with oil, which is set in the pipe.

The temperature of the escaping gases is determined in a few cases by means of a pyrometer, the stem of which is wholly exposed to the gases in the centre of the flue. In general, the temperature is obtained by means of a high grade thermometer set in an oil pot filled with cylinder oil, which is lowered into the centre of the flue. The pot is withdrawn when the thermometer is read. When a pyrometer is used, it is verified by comparisons with a thermometer.

The draught suction is determined by means of a U-tube gauge, of the form shown in the appended cut, connected to the flue between the damper and the boiler. The tube, which is made of glass, is provided with enlarged chambers at the top, arranged so as to give a magnified indication of the vacuum produced by the draught. The instrument is filled with two liquids of different color, one liquid occupying the whole of one side of the instrument, and that part of the other side to near the top of the U-tube; and the other liquid occupying the remaining space in the U-tube and the opposite chamber. When connection is made to the flue (the proper side being connected) the line of division between the two liquids, which, owing to the differences of color and character, is clearly defined, moves downward in the tube and shows by the distance traversed the magnified amount of vacuum which is acting. The amount by which the true indication is multiplied depends upon the relation which exists between the diameter of the two chambers and that of the two legs of the tube. An instrument which multi-

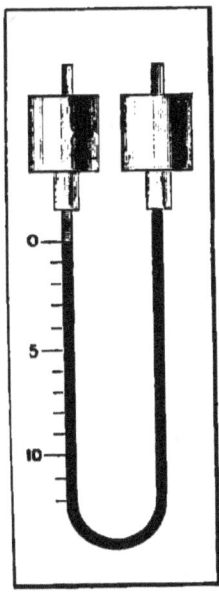

plies eight times, which is a sufficient increase, gives a movement of 4 inches for an actual draught of half an inch; that is, for a draught corresponding to the pressure of a column of water half an inch high.

The quality of the steam, in the case of boilers giving superheated steam, is determined by taking its temperature. A thermometer is inserted for this purpose in an oil cup, which is screwed into the main discharge pipe. The quantity of superheat is the number of degrees which the indicated temperature is in excess of the normal temperature. The normal temperature is that indicated by the thermometer when the boiler is under steam of the average pressure, at a time when the production has ceased, that is, when the steam is in a saturated condition.

In the case of boilers which do not superheat the steam, the quality is determined by the use of a calorimeter. The form of calorimeter employed is sometimes the barrel calorimeter, but preferably the continuous superheating calorimeter, devised by the author. The barrel calorimeter consists of a common oil barrel, fitted with an outlet valve, together with delicate scales and a finely graduated thermometer. A series of tests is made, one following after the other as rapidly as possible and the results are averaged. The barrel is first filled with water to the desired point and the water heated in preparation for the tests. This water is let out and thrown away. The barrel is then filled again and the average temperature of the incoming water is taken, read to tenths of a degree, and the weight of water drawn in is carefully determined. After blowing out the condensed water from the pipe which is provided for supplying the calorimeter, a movable piece of pipe which conducts the steam into the water is attached, and the valve is opened for the formal conduct of the test. The water is heated to about 110 degrees. Then the valve is shut, the movable pipe unscrewed, the water stirred and the temperature carefully observed. The weight is taken, and the quantity of steam condensed is found by subtracting the previous weight. Each test of the series is conducted in the same manner, except that the preliminary heating is required for none but the first test. To compute the percentage of moisture, the weight of cold water drawn in is multiplied by the number of degrees difference between the two temperatures observed, and the product is divided by the weight of condensed steam. To the

THE SUPERHEATING CALORIMETER.

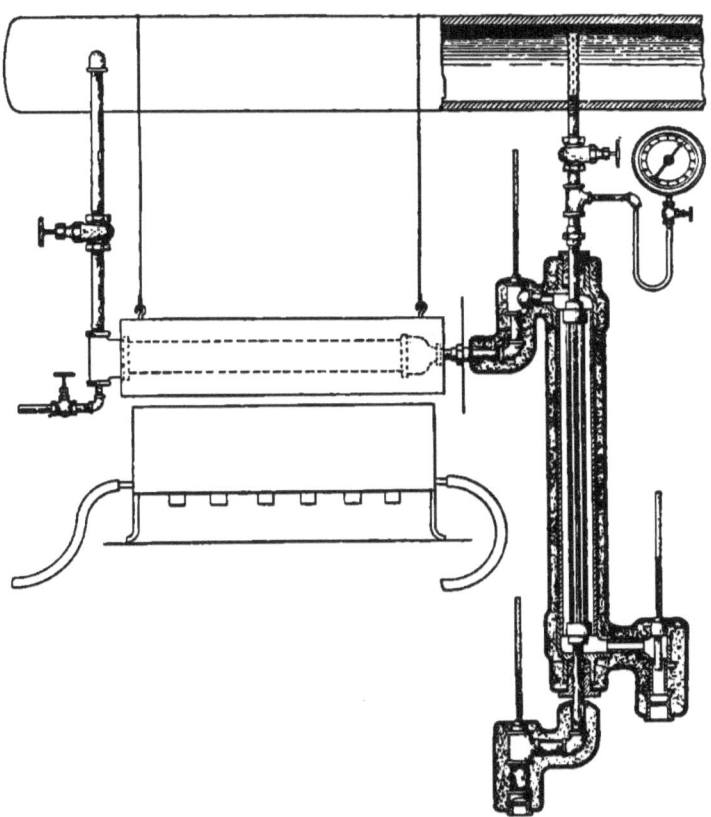

quotient is added the temperature of the heated water, and the sum thus found is subtracted from the total heat of saturated steam of the observed pressure (reckoned from zero). The remainder, divided by the latent heat of the same steam, gives the proportion of moisture.

The superheating calorimeter, which is shown in the appended cut, works on a different principle, requiring no water and no weighing. It consists of a heater, through which the steam to be tested is passed, and in which the moisture contained is evaporated. This apparatus determines the quantity of moisture, by first measuring the quantity of heat expended in evaporating it, and this is readily done by supplying the heater with steam previously superheated and observing the fall of temperature which the steam suffers in passing through the heater. The principal data required are the temperatures given by three thermometers, which do not need to be of delicate construction, nor to be read with extreme care. The quantity of moisture is readily determined to a small fraction of one per cent., since a fall of some 18 degrees in the temperature of the supply of steam to the heater is required for one per cent. of moisture.

The steam which is supplied to the calorimeter, whichever form is employed, is taken from the main discharge pipe of the boiler by means of a well covered half-inch pipe. A long thread is cut upon the pipe, and it is screwed into the main so as to extend across the whole diameter of the pipe, and the enclosed portion is perforated. In this manner the points of supply are distributed and a sample of the steam is obtained.

EXPLANATION OF THE TABLES OF PART SECOND.

Referring now to the various tables of results given, some explanation is needed as to the meaning of the lines and the method of computation employed.

When the " manner of start and stop and kind of run" is stated to be *ordinary*, reference is made to the method just described, the tests beginning early in the morning, after the boiler had stood with no fire and with closed dampers during the preceding night, and with no preliminary heating; there being an interval at noon-time when the production of steam nearly ceased, and the night evaporation being included in the total. When " manner of start and stop and kind of run " is *ordinary with preliminary heating,* the method is the same, with the exception that the night evaporation is not included in the total, and the boiler is heated for a time preparatory to the beginning of the test. This is true also when the start and stop is made with a thin fire. When the "kind of run" is stated to be *continuous,* there is no interval in the production of steam at noon-time; and when it is spoken of as a *factory* run, reference is made to the case where the usual noon-day pause occurs.

The " duration " of the test is the time required to evaporate the total quantity of water, assuming the rate of evaporation which occurs when the boiler is doing its normal work. It is the quotient obtained by dividing the total evaporation by the weight of water evaporated per hour.

The quantity of " ash " is the refuse of the furnace and ash-pits left after deducting the unconsumed coal which fails to pass through a $\frac{3}{8}$-inch screen, and the percentage is found by dividing this by the weight of dry coal consumed, and multiplying the decimal fraction thus found by 100.

The quantity of " coal consumed per hour " is found by dividing the total quantity by the number of hours determined upon for the duration of the test.

The quantity of " water evaporated per hour " is found by taking the weight evaporated during those portions of the test when the boiler is doing its normal work, say, in the case of a factory working 10 hours per day, from 7.15 to 11.45 A. M., and from 1.15 to 5.45 P. M., and dividing this quantity by the number of hours covered by those periods.

The " horse-power " is determined by first computing the equivalent evaporation per hour, supposing the feed-water is supplied at 100 degrees and evaporated at 70 pounds pressure, and then dividing this quantity by 30. This is the horse-power basis of rating established by the Committee on Boiler Trials, appointed by the American Society of Mechanical Engineers. The equivalent evaporation is found by multiplying the actual hourly rate by the total heat of the steam, reckoned from the temperature of the feed-water, and dividing the product by

1,110, which is the total heat of steam of 70 pounds pressure, reckoned from 100 degrees.

The observations of the boiler pressure, the various temperatures, and the draught suction are made every half hour while the boiler is doing normal work, and the results are averaged.

The water evaporated " per pound of coal from and at 212 degrees" is computed by multiplying the quantity evaporated per pound of coal by the total heat of the steam, reckoned from the temperature of the feed-water, and dividing the product by 966.

The water evaporated " per pound of combustible from and at 212 degrees" is computed by dividing the " water per pound of coal from and at 212 degrees," by the percentage of combustible in the coal; that is, by 100, less the percentage of ash, and multiplying the result by 100.

In the tables of dimensions, the area of heating surface given is that exposed to the fire and products of combustion, and not that exposed to the water or steam.

No attempt has been made in these tests to determine, by chemical analysis, the constituent elements in the composition of either the coal or the escaping gases.

The evaporative results expressed in pounds of water evaporated per pound of coal, include no allowance for superheating in cases where superheating exists, and are not corrected for moisture when the steam is found to be wet. The quantities given as " water evaporated" are feed-water supplied and, presumably, evaporated.

DISCUSSION OF RESULTS.

Having now considered in a brief manner the general subject of boiler tests, and pointed out the method of conducting the particular tests under notice, we may pass to the chief object of the paper; viz., a review of the various results, and a consideration of the general subject of economy in the generation of steam, as revealed by the tests.

1. COMPARISON OF BOILERS WHICH PRODUCE SATURATED STEAM WITH THOSE PRODUCING SUPERHEATED STEAM.

The boilers may be divided into two general classes, according as they furnish saturated steam or superheated steam. The first class are, as a rule, of the horizontal tubular type, and the second class of the vertical type. In most of the boilers which superheat, that is, vertical boilers, the superheat is obtained by sacrificing the water-heating surface, and providing a large area of steam-heating surface. In the horizontal tubular boiler, the proportion of heating surface to grate surface is generally 33 or more to 1, and this is almost wholly water surface. In many of the vertical boilers, the proportion of water surface to grate is only 20 to 1, that of steam surface to grate about 10 to 1, making a total of 30 to 1, and this total surface is less than that given for the horizontal boiler. Considering that the economic performance of the boiler is affected by the proportion which exists between the heating surface and grate surface, as is shown in a later part of the paper, these boilers are handicapped at the outset by a deficiency of surface, even though the steam-heating surface is considered equally as effective as the water-heating surface. Let us make a comparison between the two types of boilers, however, taking the proportions which exist. Although the evaporation per pound of coal is less in the superheating boiler than in the boiler which does not superheat, it does not follow that the actual efficiency of one is different from that of the other. Superheated steam is of greater value than saturated steam, according as it is more or less superheated. The reason for this is, first, that superheated steam contains the greater amount of heat; and, second, that heat expended in superheating steam secures a greater return, in cases at least where it is used for power, than the original heat expended in evaporating water. If, for example, steam is superheated 100 degrees, the added heat which it contains is 48 thermal units per pound, or 4.8 per cent. of the heat required to convert a pound of water, supplied at 212 degrees, into steam of 80 pounds pressure. In other words, if the boiler evaporates with one pound of coal 10 pounds of water supplied at 212 degrees into steam of 80 pounds pressure, and superheats that steam 100 degreees, the excess of heat due to its superheat corresponds to an additional evaporation of 0.48 pounds of water. When superheated steam is used for motive power, the equivalent value of the added heat is more than the quantity named. Many tests have been made which show that this is true. Comparative tests by the author show that for every ten degrees of superheating, the amount of steam consumed by an engine is reduced from 1 to 1.5 per cent. If, to be on the safe side, we take a saving of 1 per cent. of steam for 10 degrees of superheating as the proper allowance, and assume that common steam contains 1 per cent. of moisture, which, in round numbers corresponds to 20 degrees of superheat, then the equivalent evaporation corresponding to the effect of 100 degrees of superheating, in the example cited, is 1.2 lbs. of water per pound of coal.

In order that the superheating boilers may have the benefit of every advantage to which they may be entitled, let us make the comparison on the supposition that the steam is used for motive power, and that, as just noted, 10 degrees of superheating is equivalent in value to 1 per cent. increase in evaporation. The tests on the two types of boilers which can be fairly compared are those made with anthracite egg, or broken, coal, the principal results of which are given in Tables No. 1 and No. 2.

The number of tests referred to in the vertical boiler table is less than those of the horizontal table, but they may be taken as representative examples, since the comparative results obtained in several cases where a different kind of fuel was used, are, in general terms, the same.

The average proportion of heating surface to grate surface for the horizontal tubular boilers is 44.7 to 1, while the total ratio in the vertical tubular boilers is 34.7 to 1. Comparing the water-heating surface in the two cases, that in the vertical boilers is about one-half of that in the horizontal boilers. The average percentage of ash in the two cases is almost identically

TABLE No. 1.—*Horizontal Tubular Boilers, Anthracite Coal.*

Number of Boiler	Ratio of Heating Surface to Grate Surface.	Kind of Coal.	Percentage of Ash.	Coal per hour per square foot of Grate.	Temperature of Escaping Gases.	Water per lb. of Combustible from and at 212 degrees.
				Lbs.	Deg.	Lbs.
3	33.8 to 1	Lackawanna,	16.7	11.1	482	10.73
5	32.2 to 1	"	14.2	10.1	443	10.76
9	34.6 to 1	Lehigh .	9.4	14.0	349	11.24
10 av.,	35.6 to 1	"	13.4	6.7	321	11.37
11	35.5 to 1	"	15.0	5.7	—	10.48
12	42.0 to 1	Lehigh .	10.1	12.2	346	11.20
15 av.,	37.0 to 1	"	10.2	12.7	380	10.70
17	26.5 to 1	White Ash	10.1	12.9	455	9.75
22	68.0 to 1	Lehigh,	12.9	11.5	350	11.03
24	26.6 to 1	"	12.0	7.9	474	9.87
25	44.2 to 1	White Ash .	15.7	9.4	369	10.61
35	47.4 to 1	Philadelphia and Reading,	10.7	11.9	428	10.60
36	57.9 to 1	"	10.3	11.9	321	11.33
42	65.0 to 1	Lackawanna,	12.3	14.2	392	11.11
43	67.6 to 1	"	15.5	9.4	335	10.39
45	60.9 to 1	"	8.1	14.5	373	11.00
Average,	44.7 to 1		12.2	11.0	387.9	10.76

TABLE No. 2. — *Vertical Tubular Boilers, Anthracite Coal.*

NUMBER OF BOILER.	Ratio of Water-heating Surface to Grate Surface.	Ratio of Steam-heating Surface to Grate Surface.	Kind of Coal.	Percentage of Ash.	Coal per hour per square foot of Grate.	Number of Degrees of Superheating.	Temperature of Escaping Gases.	Water per lb. of Combustible from and at 212 degrees.
					Lbs.	Deg.	Deg.	Lbs.
51	20.6 to 1	9.2 to 1	Lackawanna,	8.6	11.1	90	480	9.56
53	32.3 to 1	15.6 to 1	Schuylkill,	11.6	15.4	42	478	10.13
55	20.9 to 1	9.2 to 1	Lehigh,	12.2	12.1	75	490	8.34
57	19.9 to 1	11.3 to 1	Lackawanna,	17.0	12.0	65	520	9.17
Average,	23.4 to 1	11.3 to 1		12.4	12.7	68	492	9.30

the same. The rate of combustion is slightly larger in the case of the vertical boiler. The temperature of the escaping gases is 104.1 degrees higher in the vertical boilers than in the horizontal boilers. The evaporation per pound of combustible is 15.7 per cent. larger in the horizontal boilers than in the vertical boilers, and the latter superheat the steam 68 degrees. Assuming as before that the saturated steam contains 1 per cent. of moisture, the superheating represents a gain in efficiency of 8.2 per cent., and there is left for the net superiority of the horizontal boilers 7 per cent.

A more satisfactory comparison may be made if those of the horizontal tubular boilers, which are called "double-deck," are discarded, as also those in which the heating surface is deficient. Throwing out for this reason Nos. 17, 22, 24, 42, 43, and 45, the average quantities for the remaining ten examples are as follows: ratio, 40; ash, 12.6 per cent.; rate of combustion, 10.6 pounds; temperature of escaping gases, 382 degrees; water per pound of combustible from and at 212 degrees, 10.90 pounds. Comparing the vertical boilers now with this new average, we have a difference of 110 degrees in the temperature of the gases, and 17.2 per cent. in the evaporation per pound of combustible. This makes the net superiority of the horizontal boilers 8.3 per cent.

The superior economy thus shown is evidently due in the main to the saving produced by the low temperature of the escaping gases. A difference of 110 degrees is sufficient to account for a considerable difference in the results. For example, refer to the tests on Boiler No. 46, which give the effect produced by utilizing the heat of the gases in warming the feed-water. Here a reduction of 107 degrees in the temperature of the gases secured an increase in the evaporation per pound of coal amounting to 7 per cent. The indication of temperature is not, however, according to the best view of the subject, a full measure of the extent of the loss in the waste gases. Owing to the brick setting with which the boilers are provided, and the tendency which hot brick-work has to become unsound by age, it is probable that, in ordinary use, a large quantity of unneeded air is admitted through the brickwork, not only into the furnace, but into the flue space as well, which on the one hand reduces the furnace temperature, and on the other hand lowers the temperature of the escaping gases. In this way a direct loss is produced by cooling the furnace; and further, the loss at the chimney, measured by the temperature, appears to be less than it really is.

It is clear that, according to these tests, the vertical tubular boiler in practical use, having the proportions given, is less economical than the ordinary type of horizontal return tubular boiler; and this is true notwithstanding the superior value of the superheated steam which they give, for which the most favorable allowance has been made.

There are two cases of vertical boilers in the paper, where the proportion of water-heating surface to grate surface is fully as large as in the best form of horizontal boiler, and the unfavorable influence of air leakage is prevented by the use of a fire-box. The boilers referred to are those numbered 58 and 60. The proportions of water-heating surface to grate surface are respectively 35.1 and 44.5 to 1. The temperatures of the escaping gases are respectively 423 and 427 degrees. The evaporations per pound of combustible from and at 212 degrees are both 12.29 pounds. Although these tests were made using high grades of bituminous coal, and for this reason cannot be fairly compared with those under consideration, the high character of the two results, contrasted with those referred to, brings out very clearly the fact of the inferiority of that type of vertical boiler which is deficient in heating surface and provided with a brick setting.

It would not be fair to conclude, because the superheating boilers referred to are at a disadvantage when compared with those which do not superheat, that this result must always follow. Were the loss from air-leakage remedied by the use of a fire-box, there is good reason to expect that the vertical boiler furnishing superheated steam would be as economical as the horizontal boiler furnishing saturated steam, and it is not improbable that with suitable arrangement of surface, it would in the long run take the first place.

No data are given on either type of vertical boiler to show whether a considerable degree of superheating can be obtained, and at the same time a high evaporative efficiency. It is doubtful if the practical restrictions which the construction of the boiler imposes will allow both of these objects to be realized at the same time.

It may be of interest, in passing, to note the effect produced by superheating the steam in an independent superheater. Under the head of Boiler No. 1, reference is made to a test on a boiler fitted with an apparatus for this purpose. By increasing the quantity of coal burned 17.8 per cent., this being the coal used in the superheater, a superheating of 228 degrees was obtained. On the basis which has been taken for computing the value of the superheat, that is, a gain of one per cent. for every ten degrees of superheating, this steam has a value of 24.8 per cent. over that of ordinary steam. Here is a net gain of 7 per cent. in favor of the superheated steam when used for motive power.

2. GENERAL CONDITIONS WHICH SECURE ECONOMY.

Glancing over Table No. 1, it appears that, in general, the highest results are produced where the temperature of the escaping gases is the least. An examination of this question may be made by selecting those tests in which the temperature exceeds the average, that is, 375 degrees, and comparing with those in which the temperature is less than 375 degrees, taking first those boilers in the table which are of the common horizontal tubular type. It will be remembered that all of these boilers use anthracite coal of either egg or broken size. The tests with high temperature are given in Table No. 3, and those with low temperature in Table No. 4.

The average flue temperatures in the two series are 444 degrees and 343 degrees respectively, and the difference is 101 degrees. The average evaporations are 10.40 pounds and 11.02 pounds respectively, and the lowest result corresponds to the case of the highest flue temperature. In these tests it appears, therefore, that a reduction of 101 degrees in the temperature of

TABLE NO. 3.—*Common Horizontal Tubular Boilers, Anthracite Coal, High Flue Temperature.*

Number of Boiler.	Ratio of Heating Surface to Grate.	Kind of Coal.	Percentage of Ash.	Coal per hour per square foot of Grate.	Temperature of Escaping Gases.	Water per lb. of Combustible from, and at 212 degrees.
				Lbs.	Deg.	Lbs.
3	33.8 to 1	Lackawanna,	16.7	11.1	482	10.73
5	32.2 to 1	"	14.2	10.1	443	10.76
15	37.0 to 1	Lehigh,	10.2	12.7	380	10.70
17	26.5 to 1	White Ash,	10.1	12.9	455	9.75
24	26.6 to 1	—	12.0	7.9	474	9.87
35	47.4 to 1	Philadelphia and Reading,	10.7	11.9	428	10.60
Average,	33.9 to 1		12.3	11.1	444	10.40

TABLE NO. 4.—*Anthracite Coal, Low Flue Temperature.*

Number of Boiler.	Ratio of Heating Surface to Grate.	Kind of Coal.	Percentage of Ash.	Coal per hour per square foot of Grate.	Temperature of Escaping Gases.	Water per lb. of Combustible from, and at 212 degrees.
9	34.6 to 1	Lehigh,	9.4	14.0	349	11.24
10	33.5 to 1	"	13.6	4.7	298	11.21
15	37.0 to 1	White Ash,	10.2	12.7	380	10.70
25	44.2 to 1	Philadelphia and Reading,	15.7	9.4	369	10.61
36	57.9 to 1		10.3	11.9	321	11 33
Average,	41.5 to 1		11.8	10.5	343.4	11.02

the waste gases secured an increase in the evaporation of 6 per cent. This result corresponds quite closely to the effect of lowering the temperature of the gases by means of a flue heater in the case already noted, where a reduction of 107 degrees was attended by an increase of 7 per cent. in the evaporation per pound of coal.

A similar comparison may be made on horizontal tubular boilers using Cumberland coal. Table No. 5 gives a list of these tests.

TABLE No. 5.
Common Horizontal Tubular Boilers, Cumberland Coal.

Number of Boiler.	Ratio of Heating Surface to Grate.	Percentage of Ash.	Coal per hour per square foot of Grate.	Temperature of Escaping Gases.	Water per lb. of Combustible from and at 212 degrees.
			Lbs.	Deg.	Lbs.
5	32.2 to 1	11.1	10.1	435	11.52
9	34.6 to 1	6.6	18.2	453	11.17
12	42.0 to 1	6.6	14.0	381	11.37
15 av.,	37.0 to 1	6.5	9.3	360	11.48
19	29.4 to 1	8.7	10.9	530	10.60
31	41.6 to 1	6.6	7.0	431	12.07
32	40.0 to 1	6.5	11.1	408	11.98
35	47.4 to 1	8.3	6.7	340	11.24
36	57.9 to 1	8.3	12.1	397	11.99
40	53.1 to 1	7.5	13.6	413	12.47
Average,	41.6 to 1	7.7	11.3	415	11.59

Here the average flue temperature is 415 degrees. Nos. 5, 9, 19 and 31 have temperatures exceeding 415 degrees. The average of these is 450 degrees, and the average evaporation is 11.34 pounds. The remaining boilers have temperatures below 415 degrees, the average of which is 383 degrees, and these give an average evaporation of 11.75 pounds. With 67 degrees less temperature of the escaping gases, the evaporation is higher by about 4 per cent. The difference here is less marked than in the anthracite tests, both in range of temperature and in economy, but it is in the same direction; that is, the highest evaporation is produced where the waste at the flue is the least.

The wasteful effect of a high flue temperature is exhibited by other boilers than those of the horizontal tubular class. This source of waste was shown to be the main cause of the low economy produced in those vertical boilers which are deficient in heating surface. Examples of the same effect are numerous in the case of nearly every type of boiler treated in the paper. The cast-iron sectional boilers Nos. 61 and 63 have flue temperatures of 575 degrees and 462 degrees, respectively, and evaporate at the low rate of 9.79 pounds and 9.61 pounds of water from and at 212 degrees per pound of combustible. The five water-tube boilers, referred to below, are likewise wasteful on account of the high temperature of the escaping gases. The temperatures range between 428 degrees and 540 degrees, and the evaporations between 9.68 pounds and 10.36 pounds for anthracite coal, and between 10.79 pounds and 10.98 pounds for bituminous coal, all of which are low results for their respective grades of coal.

Number of Boiler.	Kind of Coal.	Temperature of Escaping Gases.	Water per lb. of combustible from and at 212 degrees.
		Deg.	Lbs.
66	Anthracite,	540	9.68
68	Cumberland,	452	10.79
69	Anthracite,	428	10.36
70	Cambria Bituminous,	471	10.93
71	Cumberland,	523	10.98

The plain cylinder boilers Nos. 47, 48 and 49 have a flue temperature, in the most favorable case, of 567 degrees, and an evaporation of only 9.22 pounds of water from and at 212 degrees per pound of combustible, and the evaporation is reduced as the flue temperature increases. The Galloway boiler No. 50, under extremely favorable circumstances as to kind of fuel, mode of firing and general management, gave an evaporation of only 11.06 pounds, this low result being due, evidently, to the fact that the temperature of the escaping gases was at the high figure of 575 degrees.

With this accumulation of examples, no other conclusion can be drawn than that one of the vital principles underlying the attainment of economy in the generation of steam, is a low temperature of the escaping gases. What the temperature should be to secure the best results is to some extent uncertain. In the examples of horizontal tubular boilers cited, the best average results where anthracite coal is used are secured with an average temperature of 343 degrees, and where Cumberland coal is used with an average of 383 degrees. It will not be far out of the way if we consider 375 degrees as a proper limit for anthracite coal, and 415 degrees for Cumberland coal. These are named for the general case. Individual boilers may, in rare instances, give excellent economy where the waste temperature exceeds these figures, and there are two or three examples furnished in the paper where this is true. There are so many instances referred to where a boiler secures a low grade of economy with more than 375 degrees in the flue when anthracite coal is used, and more than 415 degrees when Cumberland coal is used, not only among boilers of the horizontal tubular type, but among those of all other types, that it seems reasonable to lay down these temperatures for a limit.

The relation between the heating surface and grate surface is important, and the question arises as to what that relation should be to obtain the highest efficiency. Keeping to the common horizontal boiler, let us select from the anthracite coal tests the boilers in which the ratio is below 40 to 1 and above 30 to 1, and compare with those in which the ratio is more than 40 to 1, taking, however, only those cases where the temperature of the gases is low and the rate of combustion is above 9 pounds per square foot of grate per hour. The tests used are those made on Boilers Nos. 9, 10 and 15 for the small ratios, and Nos. 12, 25, and 36 for the large ratios, and the averages of the two sets are as follows:—

Ratio of Heating Surface to Grate Surface.	Water per lb. of Combustible from and at 212 degrees.
	Lbs.
36.4 to 1	11.04
48.0 to 1	11.05

There is a difference here of 11.6 in the ratio given, and practically no difference in the character of the results. Nothing seems to be gained in these cases by increasing the surface above a ratio of 36.4 to 1, although the increase amounts to about one-third. Carrying the inquiry farther, and taking the so-called double-deck boilers, of which there are four instances given in Table No. 1, the average ratio is 65.3 to 1 and the average evaporation is 10.88 pounds. Here a loss is produced, although the surface is increased to the enormous extent of 80 per cent. These comparisons are made with different kinds of anthracite coals of large sizes, and with different arrangements of boilers, and some allowance must be made for the possible effect which variations in these conditions may have on the results. But the comparisons are suggestive. The evidence here given shows that a ratio of 36 to 1 provides a sufficient quantity of heating surface to secure the full efficiency of anthracite coal where the rate of combustion is not more than 12 pounds per square foot of grate per hour.

Individual examples are given, which furnish evidence as to the extent of surface required when bituminous coal is used. Boilers No. 28 and No. 29 are cases in point. Here an increase in the ratio from 36.8 to 42.8 secured a small improvement in the evaporation per pound of coal, and a high temperature of the escaping gases indicates that a still further increase would be beneficial. Among the high results produced on common horizontal tubular boilers using bituminous coal, the highest occurs in Boiler No. 40 where the ratio is 53.1 to 1. This boiler gave an evaporation of 12.47 pounds. The double-deck boiler, No. 42, furnishes another example of high performance, an evaporation of 12.42 pounds having been obtained with bituminous coal, and in this case the ratio is 65 to 1. These examples indicate that a much larger amount of heating surface is required for obtaining the full efficiency of bituminous coal than for boilers using anthracite coal. There is sufficient reason for this requirement in the fact that bituminous coal is of a gaseous nature, and the heat generated in its combustion is spread through a larger space. The temperature of the escaping gases in the same boiler is invariably higher when bituminous coal is used than when anthracite coal is used, and this points to the same characteristic. In practice, the deposit of soot on the surfaces when bituminous coal is used interferes with the full efficiency of the surface, and an increased area is demanded as an offset to the loss which this deposit occasions. It would seem, then, that if a ratio of 36 to 1 is sufficient for anthracite coal, from 45 to 50 should be provided when bituminous coal is burned, especially in cases like those referred to, where the rate of combustion is above 10 or 12 pounds per square foot of grate per hour.

The size of shell in horizontal tubular boilers appears to have little effect on the economy. The best of all the results with anthracite coal, which is 11.53 pounds of water from and at 212 degrees per pound of combustible, was obtained in a case where the diameter of the shell was 48 inches, and this result is all that can be expected from any boiler, whether the shell is large or small.

The number of tubes controls the ratio between the area of grate surface and area of tube opening. Boilers No. 42 and No. 45 have a very large number of tubes, and consequently a small ratio of grate to tube opening. The ratio is 5.2 to 1. They also have the very large area of heating surface represented by ratios of 65 and 60 to 1. Notwithstanding the ample provision of surface and other favorable conditions, the evaporation with anthracite coal is no higher than boilers give which have surface of much less extent, though of such character that the tube opening bears a smaller proportion to the grate surface. The conclusion which is well warranted by this fact is that a certain minimum amount of tube opening is required for efficient work. This conclusion is borne out by the result of the tests with anthracite coal on Boiler No. 12, where the products of combustion make two circuits through the shell and the ratio of grate surface to tube opening is 11.6 to 1. The ratio of heating surface to grate here is 42 to 1, and the average evaporation per pound of combustible from and at 212 degrees is 11.16 pounds. The best results obtained with anthracite coal in the common horizontal boiler are in cases where the ratio is larger than 9 to 1. From these facts the conclusion is drawn that the highest efficiency with anthracite coal is obtained when the tube opening is from one-ninth to one-tenth of the grate surface.

When bituminous coal is burned the requirements appear to be different. The effect of a large tube opening does not seem to make the extra tubes inefficient when bituminous coal is used. The highest result on any boiler of the horizontal tubular class, fired with bituminous coal, is obtained where the tube opening is largest. This is Boiler No. 40, which gives an evaporation of 12.47 pounds, and the ratio of grate surface to tube opening is 5.4 to 1. The next highest result is produced in Boiler No. 42, already alluded to, which gives 12.42 pounds, and the ratio is 5.2 to 1. Another high result is produced by Boiler No. 44. This is 12.03 pounds, and the ratio is 4.1 to 1. Table No. 5 gives three high results, averaging 12.01 pounds, and here the average ratio is 7.1 to 1. These instances are sufficient to exhibit the need of a larger area of tube opening when bituminous coal is used than when anthracite is used, and this might be expected in view of the gaseous nature of the products of combustion. Without going to extremes, the ratio evidently most to be desired when bituminous coal is used is that which gives a tube opening having an area of from one-sixth to one-seventh of the grate surface.

One set of tests is given which bears on the question as to the effect which size of tubes has upon the economy. These are the tests made on Boilers No. 28 and No. 29, in one of which 140 3-inch tubes are used, and in the other 100 $3\frac{1}{2}$-inch. The boiler with the smaller tubes gave the best result, but the improved performance was evidently due to the increased heating surface, of which there was an addition of one-sixth, rather than to any difference in the diameter of the tubes. It might be inferred from the fact that bituminous coal requires a larger collective area of tubes for best results than anthracite coal, that it may also require a larger individual area, and therefore larger diameter of tubes. This inference is not borne out by a comparison of the tests on Boilers No. 42 and No. 40, one of which had 3-inch tubes and one 3½-inch, though the two boilers are of somewhat different type. A practical objection to the use of too small tubes must be kept in mind in those cases where a very smoky grade of bituminous coal is used, and frequent opportunity cannot be had to clean the tubes, so as to prevent a serious accumulation of soot.

There appears to be no reason why the relations of heating surface to grate surface, found desirable in the horizontal boiler, should not apply with equal force to the vertical boiler, and this view of the matter is justified by the results of the test on Boiler No. 60. In this boiler there are 44.5 square feet of water-heating surface to 1 of grate, and 7.1 square feet of grate to 1 of tube area, which agree practically with the proportions named for the best work in horizontal boilers using bituminous coal. If we allow the equivalent evaporation for the effect of the 18 degrees of superheating in the same manner as in the tests on the vertical boilers, considered under a previous heading, the evaporation into saturated steam containing one per cent. of moisture from and at 212 degrees per pound of combustible in this case is 12.75 pounds.

The discussion of the general conditions which secure economy applies to medium rates of combustion of say 10 to 12 pounds per square foot of grate per hour, such as will secure the rated capacity of the boiler when the power is based on 12 square feet of water-heating surface per horse-power.

3. COMPARISON OF DIFFERENT KINDS OF BOILERS.

It seems to the author that the general principles of economy, to which attention has just been devoted, though deduced mainly from a study of horizontal tubular boilers, are based upon such reasonable grounds that they may be applied to all forms of steam boilers. The proportion most to be desired between tube opening and grate surface cannot be applied to a boiler which has no tubes; but the relations established between heating surface and grate surface, and the best condition as to the temperature of the escaping gases, are principles which can be applied in all cases. Before making a comparison of the economy of different types of boilers, these principles should be remembered, and the comparison based on the performance of the boiler when suitable conditions exist for the attainment of the best result. There is only one case where this cannot be done, and that is where the construction of the boiler forbids the attainment of those conditions.

A large proportion of the boilers treated of in the paper are, in one form or another, of the horizontal tubular type. The number of these is 46. Of this number, there are 31 of the common return tubular type; 1 is what may be termed a direct tubular with common furnace; 2 are direct tubular with detached furnace; 1 has two furnaces for alternate firing; 2 are provided with a water leg for the front of the furnace; 3 have such an arrangement of tubes that there is a double passage of the products of combustion through the boiler; 6 are of the double-deck type.

Of the remaining boilers, 3 are plain cylinder, 10 vertical tubular, 3 cast-iron sectional, 8 water-tube, and 1 Galloway.

The most favorable results with the common horizontal boiler are those obtained on No. 10 with anthracite coal, and on Nos. 31, 32, and 40 with Cumberland coal. The first is 11.37 pounds, and the average of the second 12.17 pounds. The conditions in all these cases were favorable.

The direct tubular boiler with common furnace, No. 14, cannot be compared with these on account of the different kinds of coal used; but this boiler can be compared with No. 13, which is of the common type and which used the same kind of fuel. In this comparison it stands at a disadvantage, but there seems to be a reason for it in the fact of the unfavorable proportion of heating surface to grate, which is only 27.7 to 1.

The direct tubular boilers with detached furnace, Nos. 8 and 36, cannot both be brought into the comparison, since the former used a mixture of screenings and bituminous coal. The latter, No. 36, which was tested with both anthracite and Cumberland coal, can be used. The anthracite coal gave an evaporation of 11.33 pounds, and the bituminous coal 11.99 pounds. Both of these are nearly as good results as the high-grade common boilers produced. Here the heating surface bears a ratio to grate surface of 57.9 to 1, and the ratio of grate to tube opening is 6.7 to 1. Under these circumstances, which are most favorable, a high result would be expected from any form of boiler, and the detached furnace cannot reasonably lay claim to any special advantage.

The boiler with two furnaces for alternate firing, No. 18, a system which is evidently of use only in cases where bituminous coal is burned, gave a result much below that of a common boiler. The temperature of the escaping gases was 472 degrees, which is too high for the best economy. But this unfavorable condition does not wholly account for the deficiency, the evaporation being 10.93 pounds. The evidence of this single instance is that the double furnace type of boiler is inferior to the common type.

The boilers which have a water leg for the front of the furnace, differ so little from the common boiler that neither loss nor gain, if either occurred, could with reason be attributed to this method of construction. Boiler No. 9, which is fitted in this manner, gave an evaporation with anthracite coal amounting to 11.24 pounds. This excellent result is largely due, no doubt, to the new condition of the boiler and to the favorable proportions which existed.

The boilers in which a double passage of the products of combustion occurred are Nos. 4, 12, and 38. Only one needs to be considered, viz., No. 12, the others being foreign to the discussion. Here, a favorable result is produced with anthracite coal, the evaporation being 11.20 pounds of water from and at 212 degrees per pound of combustible, and here, again, favorable conditions lent their aid. The result obtained with bituminous coal is rather low. According to the conclusions which we have reached as to the best proportions of boilers, the tube opening is insufficient, being only a little more than one-half of that laid down for bituminous coal. This form of boiler is not well adapted for securing a large tube opening, because with a given number of tubes only half of them can be employed for carrying the products in one direction. The desired end must be secured by reducing the length of the tubes and increasing their number, using a larger shell. The test gives no indication of the result which would follow this change of construction.

The double-deck boilers No. 42 and No. 45, burning anthracite coal with high rates of combustion, give an average evaporation of 11.05 pounds. Those numbered 42 and 44, burning bituminous coal, give an average of 12.22 pounds. One is slightly below the standard of the common tubular boiler and the other slightly above it. Here the benefit with bituminous coal is due undoubtedly to favorable proportions.

With all the various modifications in the type of horizontal tubular boiler to which the tests refer, some of which it must be admitted give excellent results, there is no other conclusion to be drawn than that the form of horizontal boiler, which with suitable proportions and operation can be depended upon to give the highest evaporation, is the common horizontal return tubular boiler, so widely used in New England factories.

Passing to the boilers of the other types named, the first is the plain cylinder. Little need be said of this boiler, it is of such evident inferiority as an economical generator of steam. The ratio of heating surface to grate surface is in no case above 10.9 to 1. If 36 square feet of heating surface is required for economical results, nothing can be expected from a proportion of less than one-third of this quantity, and it is probably out of the question to institute a re-arrangement in this type of boiler which will secure the desired end in any practical way.

The next form is the vertical boiler. This has already been taken up under the head of superheating boilers, and only the conclusions there derived require mention. These are, that if suitable attention is given to the design of the boiler, the vertical tubular compares favorably with the horizontal boiler. The example afforded by Boiler No. 60, using Cumberland coal, corroborates this view in the most positive manner. The result obtained in this one case, when reduced to the basis of horizontal boilers giving steam containing one per cent. of moisture, in the manner pointed out for superheating boilers, is an equivalent evaporation of 12.75 pounds of water from and at 212 degrees per pound of combustible. We have a result here which surpasses any given in the paper by horizontal tubular boilers. Vertical boiler No. 58, though of a different arrangement, gave a result which, without allowance for superheating, is precisely the same. These instances are not of sufficient number to establish the vertical boiler as one of superior economy to the horizontal boiler, but they show that, when properly designed, it is at least the equal of the horizontal boiler.

The three cases of cast-iron sectional boilers, viz., Nos. 61, 62 and 63, give comparatively poor results. The unfavorable showing is due in some degree to unfavorable conditions. In every one, the ratio of heating surface to grate surface is below that which has been given as essential to economical work. In all three cases the effect of this departure from correct principles is seen in the high degree of flue temperature. There is nothing in the construction of the boiler to prevent the employment of any proportions between heating surface and grate surface which may be desired, and it is not improbable that the results would have been equal to those of the tubular boiler if suitable proportions had existed.

There is some variety in the economic results produced by the water-tube boilers. Out of the whole number of boilers given, amounting in all to nine, only one appears to reach the standard laid down for good economy. The principal results are given in Table No. 6. Examining these figures closely, it is seen that the result referred to (Boiler No. 65,) was produced by boilers having a sufficient ratio of heating surface to grate surface, and in which there was a high rate of combustion and a low temperature of the flue. In addition to this, the boilers were comparatively new. These are conditions which, as already shown, usually lead to high performance. The inferior results given for the remaining water-tube boilers, with one exception, can be attributed to waste heat at the flue. In Boiler No. 69, the high temperature seems to be due to a deficiency in the quantity of heating surface. In most of the

TABLE No. 6. — *Water-Tube Boilers.*

Number of Boiler.	Ratio of Heating Surface to Grate.	Kind of Coal.	Percentage of Ash.	Coal per hour per square foot of Grate.	Temperature of Escaping Gases.	Water per lb. of Combustible from and at 212 degrees.
				Lbs.	Deg.	Lbs.
64	37.3 to 1	Lehigh Chestnut,	14.0	9.3	337	10.61
65	40.0 to 1	Shamokin Pea,	16.6	14.0	371	11.26
66	40.3 to 1	Lehigh Broken,	9.2	17.8	540	9.68
67	36.5 to 1	Lehigh Chestnut, No. 2,	14.7	8.2	360	10.00
68	62.5 to 1	George's Creek Cumberland,	7.7	16.8	452	10.79
69	31.4 to 1	Lackawanna Chestnut, No. 2,	16.4	10.9	428	10.36
70	45.5 to 1	Cambria,	10.5	16.9	471	10.93
71	48.4 to 1	George's Creek Cumberland,	6.4	16.3	523	10.98

remaining boilers, it appears to be inefficiency of the surface which causes loss, and this may be attributed either to deposit of scale on the interior of the surface, or of soot and kindred substances on the exterior. The last may be of frequent occurence if the care of the boiler is neglected, more especially since the exterior surfaces of water-tube boilers are difficult of access, and are usually cleaned only by the use of a jet of steam. On the whole, the water-tube boilers may be considered to be equal in economy to the tubular boilers, under the conditions of comparison determined upon, that is, when the conditions are favorable to economy.

It should be noted that one test is given on a water-tube boiler which shows an exceptionally high performance. This is referred to in Part Second, being made on Boiler No. 68. The evaporation obtained is 13.01 pounds, from and at 212 degrees per pound of combustible. The indication of this single instance shows high possibilities of the boiler, though it loses much of its force in view of the greatly inferior result obtained from the same boiler on Test No. 134, made two years later.

There remains the single Galloway boiler, which was tested with Cumberland coal. Unfavorable conditions for economy exist here, and the low result produced is no different from what would be expected. According to the principles governing economical work many times exemplified in this discussion, there were disadvantages on every hand, and the test furnishes no guide as to the capabilities of this form of boiler when operated under favorable conditions.

The general conclusion to be drawn from all these comparisons is that the economy with which different types of boilers operate depends much more upon their proportions and the conditions under which they work, than upon their type; and, moreover, that when these proportions are suitably carried out, and when the conditions are favorable, the various types of boilers give substantially the same economic result.

4. COMPARISON OF DIFFERENT KINDS OF FUEL.

There are two methods of treating the question of a comparison of different kinds of fuel. One method determines the relative economy of the various fuels, when each is burned under the conditions, regarding type and arrangement of boiler, which will give the most favorable results. The other method compares the various results obtained when the fuels are burned in the same boiler, regardless of proportions and regardless of the special adaptability of the boiler for the economical use of any particular kind of fuel. The true economy of a fuel depends to some extent upon its market price. Changes which occur from time to time in the relative market prices of different fuels, make one fuel the most economical at one time, and another fuel at another time. It becomes advisable, on that account, to change fuels when a sufficient change of prices occurs. If this is done, it is not usually practicable to make any material alteration in the boiler, and as a consequence one design of boiler must answer for all kinds of coal. It is for this reason that the second method alluded to is the one selected for the treatment of the subject, being, moreover, the one to which most of the tests conform.

A comparison is first made on the basis of the number of pounds of water evaporated from and at 212 degrees per pound of coal. Anthracite broken coal is used as a standard, and the comparison is made by determining the percentage of increase or decrease in each case above or below the evaporation with that coal. Table No. 7 gives a summary of the results of those tests which can be treated in this manner.

TABLE No. 7.

Number Designating Test.	Cumberland (increase).	Chestnut (decrease).	Pea (decrease).	Pea and Dust and Cumberland (decrease).	Pea and Dust and Culm (decrease).	Nova Scotia Culm (decrease).
	Per cent.	Per cent.	Per cent.	Per cent.	Per cent.	Per cent.
7	10	–	–	–	–	–
9	–	–	–	14	–	–
10	–	–	–	–	–	21
27	7	–	–	–	–	–
28	–	–	–	6	–	–
35	–	4	–	–	–	–
36	–	–	13	–	–	–
38	14	–	–	–	–	–
39	–	–	–	*2	–	–
50	–	–	6	–	–	–
51	–	–	–	–	14	–
72	9	–	–	–	–	–
75	8	–	–	–	–	–
76	–	–	–	*1	–	–
84	20	–	–	–	–	–
85	–	–	–	–	2	–
87	17	–	–	–	–	–
88	–	–	–	–	–	7
110	–	–	–	4	–	–
111	17	–	–	–	–	–
Average,	12.8	4	9.5	4.2	8	14

* Increase.

If we assume an evaporation of 11.00 pounds of water from and at 212 degrees per pound of combustible for the performance of anthracite broken coal, which seems to be a result within easy reach in good practice, and farther assume 11 per cent. for an average percentage of ash in such coal, the above comparison will be applied to an evaporation of 9.79 pounds of water from and at 212° degrees per pound of coal. The performance of the various fuels, expressed in pounds of water evaporated per pound of dry coal, will then be as follows:—

NAME OF COAL.	Water from and at 212 degrees per lb. of dry coal.
	Lbs.
Anthracite broken,	9.79
Cumberland,	11.04
Anthracite Chestnut,	9.40
Anthracite Pea,	8.86
Two parts Pea and Dust and one part Cumberland,	9.38
Two parts Pea and Dust and one part Culm,	9.01
Nova Scotia Culm,	8.42

These figures apply to dry coal. The fine grades of anthracite coal and the bituminous coals are frequently moist from exposure to the weather. In the moist or even wet condition which exists when the purchaser buys coal, this comparison does not indicate the true relative performance. The quantity of moisture which such coal contains is uncertain and variable. Three per cent. may be allowed for an average, but in individual cases it may run up to 6 per cent. or more. While this allowance serves to reduce the evaporation with the coals named, the coarse grades of anthracite coal in practice require an allowance which, though of different nature, operates in the same manner. The figures of the tests are based on the weight of coal obtained by deducting unconsumed fuel left at the end of the test, and no deduction of much consequence requires to be made except when the large grades of anthracite coal are used. In the practical work of operating boilers much of this unconsumed coal is thrown away with the ashes. It readily amounts to 3 per cent., and this corresponds to the allowance which with other fuels may be made for moisture. The quantities given may therefore be used, as they stand, for purposes of comparison, although taken individually they do not show precisely the work of the various fuels in the conditions in which they are bought and used.

In the comparison of relative quantity of water evaporated by the different fuels, it is seen that Cumberland coal easily takes the lead. As Table No. 7 shows, the evaporation of this coal is 12.8 per cent. more than that of anthracite coal. In the other cases the evaporation is less than that of anthracite coal; that of chestnut coal is 4 per cent. less; pea coal 9.5 per cent.; mixture of pea and dust and Cumberland 4.2 per cent.; mixture of pea and dust and culm 8 per cent.; Nova Scotia culm 14 per cent.

Proceeding farther, let us examine the subject from the financial standpoint and make a comparison on the basis of cost. Two elements must be considered here, viz., cost of fuel and cost of labor; both vary for the different fuels. Take the case of a plant of 1,000 horse-power. According to the standard of horse-power used in the paper, the daily product of steam for ten hours' run of such a plant is 344,721 pounds, assumed to be evaporated from and at 212 degrees. The number of tons of the various kinds of coal used per day, computed from this quantity, are given in Table No. 8. This table also contains the prices of the various coals, found by taking an average of the present (October, 1888,) quotations for a cargo delivered in the cities of Lawrence, Fall River and Waltham; and farther, the total cost of coal for one day's run at the prices named. The table also gives the cost of labor required in firing, the assumption being made in the case of anthracite coal of broken, chestnut or pea sizes, that the labor is performed by two day firemen, one night fireman and two helpers, and in the case of bituminous coals that one additional fireman is required; while in that of the mixed fuels, one fireman and one helper additional are required. The wages of the firemen are assumed to be $1.75 per day, and of helpers $1.25 per day. Finally, the table gives the total cost of fuel and labor for the day's run of ten hours, computed in accordance with the assumptions named.

It appears from Table No. 8 that at the prices named there is a difference in the cost of coal and labor for a day's run on a 1,000 horse-power plant, reckoned between the highest and lowest quantities, of $41.27. The significance of this sum is apparent when it is considered that in a year's time it represents $12,711. Such a sum has no small relation to the total amount of profit which a mill employing a 1,000 horse-power plant realizes.

COMPARISON OF DIFFERENT KINDS OF FUEL. 47

TABLE No. 8.— *Cost of Coal and Labor for a Day's Run of Ten Hours, 1,000 Horse-power Plant.*

	Anthracite broken.	Cumberland.	Anthracite Chestnut.	Anthracite Pea.	Two parts Pea and Dust and one part Cumberland.	Two parts Pea and Dust and one part Culm.	Nova Scotia Culm.
Weight of coal used in ten hours (2,240 lbs. = 1 ton), . . . tons	15.7	13.9	16.4	17.4	16.4	17.1	18.3
Cost of coal per ton of 2,240 lbs., . $	5.65	4.56	6.13	3.74	3.72*	3.29	3.28
Cost of coal used in ten ho_rs, . $	88.70	63.38	100.53	65.25	61.05	56.26	60.02
Cost of labor per day, . . $	7.75	9.50	7.75	7.75	10.75	10.75	9.50
Total cost of coal and labor per day, $	96.45	72.88	108.28	73.00	71.80	67.01	69.52

* Price of pea and dust $3.30 per ton.

The highest cost in the comparison attends the use of anthracite chestnut coal, and the lowest that of a mixture of two parts pea and dust and one part Nova Scotia culm. The pure Nova Scotia culm stands about on a par with the mixture.

If the Nova Scotia coal and the mixture containing Nova Scotia coal are left out of the comparison, the mixture of pea and dust and Cumberland coal takes the lead as the cheapest fuel, and the Cumberland follows closely upon it. These figures show that at prices which now hold, little is gained by the use of a mixture of either pea and dust and Cumberland or of pea and dust and culm. Either of the bituminous coals fired without mixture produces nearly as good results. Considering the fact that the use of the mixed fuel reduces the available power of the boiler, the pure coal easily has the advantage; no one would go to the trouble of using an inferior grade of coal, such as the best pea and dust mixed with bituminous coal is, and such as much of it is liable to be, unless by so doing a material reduction in cost could be attained. If yard screenings can be delivered at $2.75 per ton, and if it secures the same result as pea and dust coal, the total cost of fuel and labor when a mixture of two parts screenings and one part Cumberland coal is used amounts to $65.69 per day, and this sum is lower than any quantity given in the table.

Pea coal comes next to Cumberland in the order of cheapness, though the difference is trifling, and then the anthracite broken coal.

An important matter relating to the use of different kinds of coal is the relative amount of power which they give when fired in the same boiler. Boiler No. 5; rated at 54 horse-power (on the basis of 12 square feet of heating surface), gave 60 horse-power with Cumberland coal, 53.9 horse-power with anthracite broken, and 38.5 horse-power with a mixture of two parts pea and dust and one part Cumberland, the damper being wide open in every case. Boiler No. 9, rated at 74 horse-power with 3-8 inch draught suction, gave 143.8 horse-power with Cumberland coal, 105.5 horse-power with anthracite egg, and 95.1 horse-power with a mixture of equal parts of screenings and Cumberland coal. Boiler No. 12, rated at 87 horse-power, gave 84 horse-power with anthracite broken coal, 105.4 horse-power with Cumberland, and 82.2 horse-power with a mixture of two parts screenings and one part Cumberland. The draught was 0.28 inches. Boiler No. 17, rated at 129 horse-power, gave 192.3 horse-power with anthracite broken coal, draught 0.11 inches; 149.2 horse-power with pea coal, draught 0.12 inches; and 157.1 horse-power with 44 parts pea and dust and 37 parts culm, draught 0.28 inches. Vertical boiler No. 54, rated at 140 horse-power, gave 94 horse-power with Cumberland coal, draught 0.07 inches; 103.8 horse-power with anthracite broken, draught 0.05 inches; and 118.1 horse-power with two parts screenings and one part Cumberland, draught 0.32 inches. Boiler No. 36, rated at 270 horse-power, gave 196.1 horse-power with anthracite broken coal, draught 0.25 inches; 214.6 horse-power with Cumberland coal, draught 0.20 inches; and 204.8 horse-power with 6 parts screenings and 4 parts Cumberland, draught 0.31 inches. It is seen, therefore, that as regards capacity, Cumberland coal secures the largest power with a given draught, and pea coal or mixtures of screenings and bituminous coal the least. This suggests that to secure the same capacity from a given boiler with different kinds of coal, the area of grate surface must be varied to suit the varied kinds of coal. In the case of Boiler No. 9, for example, the grate would have been increased in the proportion of 95.1 to 143.8 when the mixed fuel was used, in order to bring the amount of power developed up to that obtained with Cumberland coal.

Among the tests given in the paper are a few made with other kinds of fuel than those upon which the comparisons are made.

There are several with "chestnut No. 2" anthracite, which corresponds in size to the pea grade. The evaporation, and the cost also, are about the same as with pea coal.

Two tests are given where Walston bituminous coal was

used. These were made on boilers No. 34 and No. 39, and the conclusion drawn from the results is that this coal under favorable circumstances gives about the same evaporation per pound of fuel as anthracite broken coal.

From the indication of a single test, made on Boiler No. 37, the performance of Ohio lump coal with favorable conditions appears to fall about 5 per cent. below the economy of anthracite broken coal.

Two tests are given which show the performance of coke, that is, the refuse coke of gas retorts used in the manufacture of illuminating gas. Test No. 95, made with coke on a plain cylinder boiler, gave 8 per cent. less evaporation per pound of fuel than anthracite pea. Test No. 67, made with coke on a horizontal tubular boiler, gave 14.7 per cent. less evaporation per pound of fuel than a high grade of Cumberland coal. The cost of the coke was $3.00 for 2,000 pounds.

Three tests give information upon the use of petroleum for fuel. Test No. 24, made with petroleum, under somewhat unfavorable conditions, gave an evaporation of 11.96 pounds of water from and at 212 per pound of petroleum. Test No. 77 with "residuum" of petroleum gave 13.66 pounds. A test with Canadian oil gave 15.00 pounds. Suppose the last figure can be realized in good practice. At this rate, the quantity of oil required to produce 1,000 horse-power for a day's run of ten hours (344,721 pounds from and at 212 degrees) is 22,981 pounds, or 3,536 gallons (1 gallon = 6.5 pounds). The price of oil per gallon, required to make the cost for a day's run equal to that of, say Cumberland coal, may be figured from Table No. 8. The total cost of fuel and labor given for Cumberland coal (the price being $4.56 per ton) is $72.88. The cost of labor when oil is used is reduced to, say one man at $2.00 per day and a helper at $1.25, making a total of $3.25. Substract this from $72.88 and there is left for the cost of fuel $69.63, which is $1\frac{97}{100}$ cents per gallon of petroleum, or 98.5 cents per barrel of 50 gallons. This means, in round numbers, that the price of oil must be less than one dollar per barrel, delivered at the boiler, in order that the cost of fuel and labor for a 1,000 horse-power plant shall be equal to that which obtains when Cumberland coal is used at $4.56 per ton.

If the economic result with petroleum goes up to 18 pounds, the price of oil per barrel to equal Cumberland coal on this basis is $1.18 cents. It is to be noted that these figures are based upon a barrel of 50 gallons. If the barrel is taken to be 42 gallons, the prices become 82.7 cents for an evaporation of 15 pounds, and 99.1 cents for an evaporation of 18 pounds.

5. MISCELLANEOUS DISCUSSION.

The general subject of boiler economy has now been examined, so far as it is affected, first, by the character of the steam, that is, as to its being saturated or superheated; second, by the general arrangement of the heating surface and the conditions under which the boiler is worked; third, by the type of the boiler; and fourth, by the character of the fuel burned. These may be regarded as the more noteworthy divisions of the subject. Incidental matters remain to be taken up, and these are quite as interesting, and sometimes almost as important, as the main questions thus far considered.

An important question connected with the economy of boilers is that of the effect which the form of setting has, where the boiler is externally fired. Questions arise in the setting of horizontal tubular boilers as to the distance between the grate and the shell, and between the top of the bridge wall and the shell; the shape of bridge wall, whether curved upward to conform to the curve of the shell, or simply left flat; the arrangement of the front of the bridge wall, whether inclined or vertical; and the arrangement of the space behind the bridge wall. Although the tests furnish no direct comparative data on these subjects, useful information may be drawn from individual results. It is significant that the highest results given are in some cases obtained with one form of setting and in some with another form. Boiler No. 9, which gave a high result with anthracite coal, has a flat bridge wall with perpendicular front, and the space behind the bridge is partially filled. Boiler No. 10, which also gave a high result with anthracite coal, has a curved bridge wall with vertical front, and the space behind is filled so as to conform to the same curve. Boiler No. 12 is another case in point, and here the bridge wall is flat with a vertical front, and the space behind is filled in so as to provide an inclined bed leading down to a deep chamber at the rear end. Boiler No. 35, which gave a good result, if allowance is made for the high temperature of the gases, has a flat bridge wall with perpendicular front, and the space behind is filled to an even level with the top. Boilers No. 13 and No. 16, using inferior grades of anthracite coal, showed favorable results, and the bridge walls are flat with vertical fronts and the space in the rear is open in both cases. Passing to boilers using Cumberland coal, No. 31 and No. 32, which give high results, both have vertical walls; No. 31 is curved at the top, with open space behind, and No. 32 is flat with open space, the latter being provided with a second wall at the rear end of the boiler. It would appear from these examples that the general form of the setting in respect to the particulars named may be one thing or another, and the boiler still give the highest economy. This furnishes strong ground for the conclusion that the matter is of comparatively little importance, and that the plan to be followed is the one which will secure the simplest, and at the same time the most convenient arrangement.

The tests give some very pointed indications regarding the effect of admitting air into the furnace above the fire. Many of the settings are arranged with a view to this special object. A current of air in a finely divided state is introduced into the furnace so as to mingle with the products of combustion as they emerge from the burning coal.

The simplest method followed is to conduct the air directly from the outside to the interior of the bridge wall, which is made hollow, and to discharge it through perforations in the top covering of the wall, which may be either iron or brick. The air thus supplied mingles with the lower strata of burning gas as it skims over the bridge. Boiler No. 15 was arranged according to this method, and tests Nos. 40 to 45 were conducted with a view to determining the effect which the admission of air has upon the economy of the boiler, three different kinds of coal being used. The coals were Cumberland, anthracite broken and a mixture of two parts pea and dust and one part Cumberland. The air was supplied to the bridge wall through an opening having an area of 38.5 square inches, which is 1.4 square inches for one square foot of grate surface. When Cumberland coal was used the full opening of this area was employed; when anthracite coal and the mixture were used the opening was contracted about one-half, thus presenting an area of 0.7 square inches for one square foot of grate. There is a wide difference in the effect of admitting air in the different cases. In the case of Cumberland coal, the evaporation is increased 5.9 per cent. per pound of coal, and 6.2 per cent. per pound of combustible; that of anthracite coal is decreased one per cent. per pound of combustible and is about the same per pound of coal; that of the mixed fuel is decreased 2 per cent. per pound of coal and 4.7 per cent. per pound of combustible. The effect which the introduction of air had upon the appearance of the products of combustion, as viewed by an observer at the " peek hole " back of the bridge wall, was very noticeable in every case. It was greatest, to be sure, with Cumberland coal, but the heightened color and the increased activity of the flame produced by the entrance of air was plainly to be seen whichever fuel was used.

It is to be noted that an apparent benefit was realized in the case of both the anthracite coal and the mixed fuel when judged by the improved state of combustion which the eye beheld. In reality, however, there was a loss, and this goes to show that the ocular method of judging in these matters is liable to be deceptive.

In these tests another effect was produced by the admission of air. The quantity and density of the smoke issuing from the chimney was reduced. This was most marked with Cumberland coal, which gave the most smoke. When air was excluded, smoke was visible $\frac{74}{100}$ of the time; when air was admitted, the smoke was visible $\frac{70}{100}$ of the time. In the case of the mixed fuel little smoke was produced in either case.

A second method of admitting air which, though not so simple, is rather more effective than that just named, is the one applied to Boiler No. 9. Here the air is supplied first to a pipe laid in the bridge wall, and then to perforated cast-iron globes which rest upon the top of the wall. This method is rendered more effective than the first by the more thorough mixture which it secures between the incoming air and the burning gases. The quantity of air supplied is increased above that naturally drawn in by means of a jet of steam. The steam thus supplied mingles with the air. In this case the effect of admitting air above the fuel, when Cumberland coal was burned, was to increase the evaporation per pound of coal 8.4 per cent., and per pound of combustible 8 per cent. It may be added that a test made with Lehigh coal, not recorded, showed that with this fuel the admission of air was attended by an increase of evaporation of 1.9 per cent. per pound of coal and 3.7 per cent per pound of combustible.

A third method of admitting air, which is more efficient than either of the first two, is that applied to Boiler No. 40. It consists in supplying one current of air through the bridge wall by means of perforations in the rear face of the wall near the top, and another current of air through a secondary wall placed a short distance to the rear of the bridge wall, and elevated so as to close up the passage between it and the shell of the boiler, and make the products of combustion descend and pass beneath it. The second current of air discharges through perforations in the front face of the hanging wall. By these provisions the entering air is thoroughly diffused through the whole volume of burning gas. The only test made on this boiler was with the air admitted, and there is consequently no positive information as to the economy of this method. There is indirect evidence, however, pointing to its superiority, in the fact that the result of the test was most excellent, being 12.47 pounds of water from and at 212 degrees per pound of combustible, and this is the highest given in the paper for horizontal tubular boilers.

A fourth method consists in first passing the air back and forth through passages running lengthwise of the walls of the setting and then discharging it through perforations, part of which are located on the top of the bridge wall and part on the two sides of the furnace. The object sought is to supply the air in a somewhat heated condition. Several tests are given which show the effect of this method. Tests No. 11 and No. 12 were made for this purpose. A mixture of two parts pea and dust and one part Nova Scotia culm was used, and in one test air was supplied through the registers in the fire-doors, and in the other through the passages in the walls. The use of the air passages secured an increased evaporation of 2 per cent. per pound of coal, while the result, when figured on combustible, was the same in both cases. Vertical boiler No. 54 is provided with air passages running up and down in the circular wall of the setting, and discharging at numerous points over the fire through perforations. Test No. 109, made on this boiler with air admitted, gave 4.3 per cent. less evaporation per pound of coal and 2.3 per cent. less per pound of combustible, than test No. 110, made with air excluded. The fuel was a mixture of two parts anthracite screenings and one part Cumberland coal. Boiler No. 20, in which air was admitted in accordance with this method, compared with an exactly similar boiler (No. 21) in which air was not admitted, fired with a mixture of three parts pea and dust and one part Cumberland coal, gave 4.5 per cent. less evaporation per pound of coal and 4.7 per cent. less per pound of combustible. Boiler No. 27, with air admitted and fired with Nova Scotia coal, gave 1 per cent. higher evaporation per pound of coal and 1.5 per cent. higher evaporation per pound of combustible than Boiler No. 26, which was precisely similar, except that no provisions were made for admitting air through the walls. In both of these cases the registers in the fire-doors were open. They presented an open area of 22 square inches, or one-fourth of one square inch for one square foot of grate surface. The openings to the passages in the walls presented an area of 32 square inches, or about four-tenths of one square inch for one square foot of grate. This proportion is only about one-fourth of that provided in the case first noticed (Boiler No. 15), where the admission of air to Cumberland coal gave about 6 per cent. advantage.

The conclusion drawn from these examples is that a considerable advantage attends the admission of air above the fuel when bituminous coal is employed, the amount of gain depending somewhat upon the method employed. There is no advantage in the system when mixtures of anthracite screenings and bituminous coal are used, if carried out according to either the first or fourth methods; and, finally, little or no benefit is derived when anthracite coal is burned.

FLUE HEATERS.

Another question connected with the general subject of the boiler economy, is that of the economy produced by the employment of a feed-water heater in the flue, and the tests show that this may be of considerable importance. That the use of a flue heater, in connection with a boiler having too little heating surface, and sending a high degree of waste heat into the chimney, is productive of economy, no one can have good reason to doubt; and this is clearly shown by the results of tests No. 121 and No. 122, which are cases in point, where a temperature of 618 degrees existed, and the heater added 29 per cent. to the evaporation per pound of coal. The special interest in the question is not in cases like this, where the waste heat is due to defects in the design and operation of the boiler, but rather in cases of reasonably good practice, where the temperature of the flue is not what may be regarded as excessive. There are a number of such cases given, and the general results produced are repeated in Table No. 9.

TABLE NO. 9. — *Tests with Flue Heaters.*

1. Number of boiler,	33	46	62	68
2. Area of heating surface, boiler, sq. ft.	1,894	4,058	5,592	8,126
3. Area of heating surface, heater, sq. ft.	1,600	1,920	1,280	1,600
4. Temperature of gases leaving boiler, deg.	376	361	403	435
5. Temperature of gases leaving heater, deg.	231	254	299	279
6. Temperature of feed-water entering heater, deg.	95	79	111	84
7. Temperature of feed-water entering boiler, deg.	175	145	169	196
8. Increased evaporation produced by heater, per cent.	10.5	7	9.3	12.8

The average results of these four trials show that the use of a flue heater having 44 per cent. as much heating surface as that of the boiler, applied where the temperature of the escaping gases is 394 degrees and the initial temperature of the feed-water 92 degrees, increases the evaporation per pound of coal 9.9 per cent. The important question which arises is, does this amount of gain pay for the addition to the plant involved by the employment of this apparatus? Let us refer the matter to a plant of 1,000 horse-power, and use for a calculation the cost of Cumberland coal required for a day's run of 10 hours, given in Table No. 8, which is $63.38. The saving produced by the flue heater is $5.71 per day, and this amount represents for a year of 308 days a total saving of $1,759. The cost of a heater with complete equipment of setting and appurtenances, having a surface of say 5,000 square feet, which corresponds to the instances referred to, is $7,000 to $8,000. The economy secured is, therefore, sufficient to pay a yearly return of from 20 per cent. to 25 per cent. on the additional investment, which is certainly of some importance.

It must be borne in mind that this result applies only to a case where the temperature of the water supplied to the heater is comparatively low, corresponding to that of the overflow water of a condensing engine. On the other hand, it is a case where the temperature of the flue is not far from that laid down in another part of the paper for boilers giving most economical results, and it is seen from the many instances recorded that the exigencies of boiler work often bring about a much higher flue temperature. The flue heater thus has a useful place, not only in a few instances but in a large majority of boiler plants where the water is supplied in a comparatively cold state. In the cases given, there is a reduction of 128 degrees in the temperature of the gases, and an increase of 79 degrees in the temperature of the water. Each one per cent. added to the evaporation corresponds to a reduction of about 13 degrees in the temperature of the escaping gases.

It is important to note the effect which the use of the heater has upon the draught. The amount of draught suction which existed in some of the above examples and in one additional example, together with information regarding quantity of coal burned and other matters bearing on the question, are given in Table No. 10.

The effect of a reduced temperature of the escaping gases, which always accompanies the employment of the heater, is seen, in every one of these instances, to reduce the draught power of the chimney. In Boiler No. 59 the temperature is reduced from 645 degrees to 365 degrees, and the draught from $\frac{41}{100}$ of an inch to $\frac{32}{100}$ of an inch, or 22 per cent. This is with an 80-foot chimney. In the last case a reduction of temperature from 575 degrees to 373 degrees is accompanied by a reduction of draught from $\frac{97}{100}$ of an inch to $\frac{69}{100}$ of an inch, or 29 per cent. This is with a 140-foot chimney. Although the general effect of the flue heater in this respect is unfavorable, a larger amount of draught is required to operate the boiler, in every instance where the average draught is given, when the heater is thrown out of use. This is not due to any peculiar action of the heater but evidently to the increased quantity of coal which must be burned in order to make up for its absence. It may be concluded that a plant of boilers working to the full capacity of the chimney without a heater, will have no difficulty in producing the same amount of steam,

FLUE HEATERS.

TABLE No. 10.— Influence of Flue Heater on the Draught.

	33		46		59		68		140	
1. Number of boiler,										
2. Height of chimney, ft.	110		80		80		80			
3. Conditions,	Heater used.	Heater not used.	Heater used.	Heater not used.	Heater used.	Heater not used.	Heater used.	Heater not used.	Heater used.	Heater not used.
4. Coal per hour, lbs.	491.7	547	852.7	875.6	683.1	894.1	760.3	843	2,952	3,282
5. Temperature gases entering chimney, deg.	231	399	254	342	365	645	279	452	373	575
6. Average draught suction inside damper, in.	—	—	.17	.22	.09	Not taken.	.25	.27	.64	.67
7. Draught suction with damper wide open, in.	.40	.51	.25	.27	.32	.41	.26	.30	.69	.97

working with a heater, unless the temperature of the gases is excessive. If the temperature is excessive and the full capacity of the chimney is required, there is some doubt whether a heater could successfully be used.

The quantities given in Table No. 10 show the draught power of chimneys of different heights, supplied with gas of various temperatures. In this connection it may be added that the draught of another chimney having a height of 230 feet, supplied with gas at a temperature of 500 degrees, amounted to 1.6 inches. This and all the observations referred to were made in cold weather.

The effect of changing the proportions of air space in the grates is shown by the tests on Boiler No. 10. With anthracite Lehigh coal, of which 8.7 pounds were consumed per square foot of grate per hour, the grates having 50 per cent. air space, secured about 2 per cent. more evaporation per pound of combustible than those having 60 per cent. air space. With about one-half as much coal burned in a given time the gain was 8 per cent. in favor of the smaller air space. The results of repeated tests corroborated these figures. Considering that whatever air passes through the grate, beyond that needed for chemically perfect combustion, produces a loss of economy, and that an excess of air to a greater or less extent always attends combustion, the effect may be explained on the ground that the reduced amount of opening in the grates tends to cut off this source of waste. The effect appears to be greatest when the rate of combustion is low. In the case mentioned the rate was much below that required to work the boiler to its nominal capacity. The effect is small at the higher rate and the conclusion may be drawn that it is only under special conditions of work that the matter is of much importance.

The tests on Boiler No. 10 furnish a case of economy produced by automatic regulation of the draught over hand regulation. The automatic regulation secured an increased evaporation amounting to 3.7 per cent. The hand regulation consisted in alternately opening the damper wide and changing it to a nearly closed position, so as to secure regular variations of pressure over a uniform range. These tests show the tendency to loss produced by an unsteady draught. They do not show, however, what may happen in the ordinary work of operating boilers. In the case given, the firing was good for both methods of regulation. In ordinary work variations occur not only in firing but in other respects, and the economy of automatic regulation may be greater or even less than that obtained in the special case cited. Owing to the important part which the personal element of the fireman plays there must be a large difference in the result produced in different cases, and the tests on some large plants, referred to in the memoranda relating to Boiler No. 10, one of which gave a result favoring automatic regulation, and the other favoring hand regulation, show this to be true.

The use of an artificial draught, made by a fan supplying warm air to the ash pit of Boiler No. 8, produced a result which was not materially different from that obtained with a natural supply of unheated air. The fuel was a mixture of anthracite screenings and Cumberland coal. There was the slight gain of 2 per cent. figured on the weight of coal used, and less than 1 per cent. figured on combustible.

The addition of heating surface to Boiler No. 2, by introducing pipes beneath the shell through which the feed-water passes before it enters the boiler, seemed to increase the evaporation per pound of coal 3.5 per cent. This gain does not count for much, because at the time when the pipes were introduced other changes in the line of improvement were made, and this percentage simply represents the collective result. It is noted in these tests that the temperature of the escaping gases is low, being in the first test only 355 degrees. Little benefit can be expected from an increase of furnace surface in a case like this, where there is an indication, in the small amount of waste heat, that the heating surface already provided absorbs about all the heat that can be given up to the boiler.

The tests on Boiler No. 43 furnish an instance where the

wetting of Cumberland bituminous coal secured an appreciable advantage. By adding 5 per cent of its weight of water, the evaporation, based on the dry weight, was increased 3 per cent.

According to the results of tests No. 97 and No. 98, the loss produced by banking a Cumberland coal fire was about 2 per cent.

The prevention of smoke was attained in an almost complete manner in Boiler No. 30, burning Cumberland bituminous coal, by the use of a sufficient quantity of air above the fuel and of superheated steam below the grate. The general economic result was, however, inferior to that obtained by boilers in which the smoke is not suppressed.

The indication of the calorimeter trials made in connection with the tests, so far as they have been conducted, is that the steam produced by the boilers which do not superheat is in nearly a dry state. Horizontal tubular boilers Nos. 2, 9, 32, 35, 36, 39, 42 and 45, gave steam containing respectively 0.6, 0.3, 2.2, 0.7, 0.5, 0.2, 0.5 and 0.3 per cent. of moisture, the average being 0.7 per cent. Water-tube boilers Nos. 65, 68, 70 and 71, gave respectively 0.6, 1.3, 0.4 and 0.5 per cent., averaging 0.7 per cent. The Galloway boiler No. 50 gave 7 per cent., and the single boiler which was tested gave dry steam.

The remaining tests would be of greater value if the quality of the steam had been obtained. It is not unreasonable to apply the general results of a number of calorimeter tests on a given type of boiler, to one of the same type where the quality of the steam is not known, but it is somewhat unsatisfactory when it is seen that boilers of one type, the horizontal type for example, give percentages of moisture varying from 0.2 to 2.2 per cent. The question naturally arises in a case where a high evaporation is produced, whether it is not due in a measure to the wet quality of the steam. This question cannot be definitely answered unless a calorimeter trial is made. In the absence of such a trial, however, there are some external evidences which throw light upon the subject. If the steam is wet and it is used in an engine, the observant engine driver will be aware of the fact by the unusual presence of water in the cylinder. The steam which escapes from the indicator cock, when it is opened, will show an excessive amount of moisture. The water which is discharged by leaking flange joints, or by the leaking stuffing box of a valve in the steam-pipe, will clearly point to wet steam. The moisture will be seen at the vent of the safety valve, if notice is taken of the escaping steam when the valve is opened. These indications can safely be credited when there is more than 3 per cent. of moisture. In regard to the application of this matter to the tests given in the paper, it may be said that there were none of these external evidences of wet steam in any case where the calorimeter trial was not made.

Little has been said here upon the subject of imperfect combustion. It has been incidentally mentioned in the treatment of boiler settings, where the effect of a supply of air above the fuel is considered. That imperfect combustion is a factor in the burning of bituminous coal, the results produced by the admission or exclusion of air clearly prove. That it is also a factor in the burning of other kinds of coal, no one will question. How far it has operated to affect the results of the various tests given, is a matter about which the reader must form his own opinion. The feeling of the writer is that imperfect combustion, which doubtless existed to a greater or less extent in all cases, is not sufficient to change the general conclusions thus far set forth. It is highly probable that a full knowledge of the character of the combustion, as determined by analysis of the flue gases in each case, would have furnished a reason for a few individual results, which, in the light of the data at hand, appear somewhat obscure. Doubtless, a knowledge of the quality of the various fuels, obtained either by chemical analysis or by determining the total heat of combustion, would also have been useful for the same end. The determination of these questions involves labor and expense which few care to undertake, and on this account it is outside the province of tests of limited scope, like those treated of in the paper.

In conclusion, it should be stated that the various boilers and appliances to which the paper refers are not designated by their commercial names, and no mention is made of the locality where they are in use. The desire to give prominence to principles rather than to individual methods of carrying out the principles, and to remove from the discussion all matters bearing upon purposes of trade, has led to this course. Furthermore, it was on condition that the tests should be treated in this impersonal way that the interested parties, with a single exception, granted permission for the public use of the results. The exception is the case of the tests on boilers Nos. 64, 65 66 and 71, which are of the Babcock and Wilcox type. These are given with the permission of the manufacturers of the boilers, on behalf of whom the tests were made.

PART II.

PART II.

Boiler No. 1.

Kind of boilers,	{ One horizontal return tubular. { One horizontal direct tubular.
Number used,	Two.
Horse-power (collective, basis 12 sq. ft.),	One hundred and seven.
Kind of coal,	Anthracite, Hazelton, Chestnut.
Age,	Fifteen years.

Boiler No. 1 consists of two horizontal tubular boilers, set side by side, as shown in the following cuts. They may be called superheating boilers, since all the steam generated by them passes through superheating pipes located in the combustion chamber under the return tubular boiler. The direct tubular boiler is provided with steam heating surface, which consists of the upper surface of the shell, a part of which is exposed to the heat of the escaping gases. This surface had not been cleaned for a long time, and probably had little effect in drying the steam.

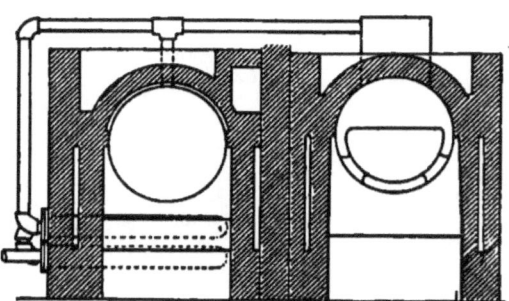

BOILER NO. 1, CROSS SECTION THROUGH FURNACES.

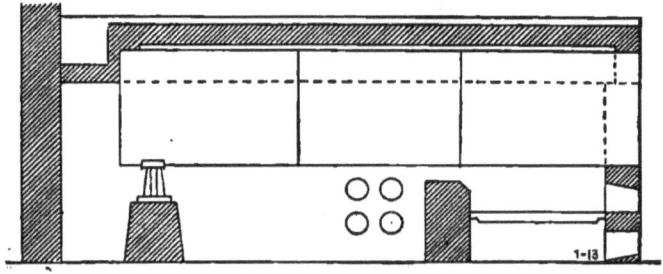

BOILER NO. 1 (*a*), LONGITUDINAL SECTION.

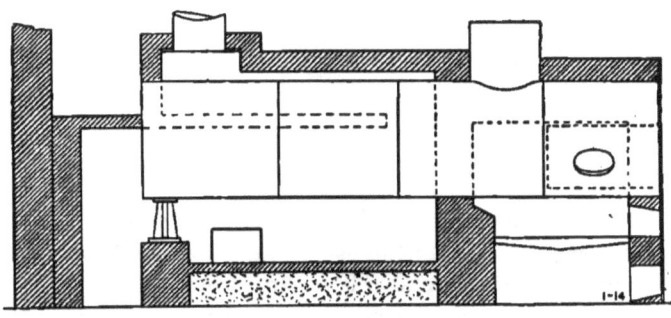

BOILER NO. 1 (*b*), LONGITUDINAL SECTION.

Dimensions of Boiler No. 1.

	Return Tubular.	Direct Tubular.
Diameter of each shell, . . . in.	48	48
Length between heads and length of tubes, ft.	16	14
Number of tubes 3 inches outside diameter,	48	45
Length of combustion chamber, . in.	–	38
Area of water-heating surface, sq. ft.	669	586
Area of steam-heating surface, sq. ft.	50	43
Area of grate surface, . . sq. ft.	19.9	21.2
Area through tubes, . . . sq. ft.	1.98	1.86
Width of air spaces and metal bars in grates, ... in.	Air 3-8 Metal 5-8	Air 3-8 Metal 5-8
Distance of grate to shell, . . . in.	18	17
Distance of bridge wall to shell, . in.	7	–
Ratio of water-heating surface to grate, .	33.7 to 1	27.7 to 1
Ratio of grate to tube area, . . .	10 to 1	11.4 to 1

Results of Tests. Boiler No. 1.

	Test No. 1.	Test No. 2.
Manner of start and stop and kind of run,	Ordinary.	Ordinary.
Duration, ... hrs.	10	9.67
Coal consumed (including wood for starting), ... lbs.	3,163	3,374
Percentage of ash, . . per cent.	15.6	–
Water evaporated, . . . lbs.	26,187	29,562
Coal per hour, ... lbs.	316.3	349
Coal per hour per square foot grate, lbs.	7.7	8.5
Water per hour. . . . . lbs.	2,618.7	3,057.1
Water per hour per square foot water-heating surface, ... lbs.	2.09	2.44
Horse-power developed, . . .	79.4	92.7
Boiler pressure, ... lbs.	73	71.6
Temperature of feed-water, . deg.	200	200
Temperature of escaping gases, . deg.	360	367
Degrees of superheating, . . deg.	66	None.
Water per pound of coal, . . lbs.	8.28	8.76
Water per pound of coal from and at 212 degrees, ... lbs.	8.66	9.16
Water per pound of combustible from and 212 degrees, ... lbs.	10.25	10.85

The object of the tests on Boiler No. 1, was principally to determine the effect which the use of the superheater had upon the economy of the boilers. Test No. 1 was made with the steam from the two boilers passing through the superheating pipes, and test No. 2 with the superheater shut off. The steam was superheated on test No. 1 66 degrees, and a comparison of the two evaporative results shows that this was attended by a loss in evaporation of $8.76 - 8.28 = 0.48$ lb., or 5.5 per cent.

In the same series of tests to which these belong, the effect of placing an auxiliary furnace beneath the superheating pipes, using coal at the rate of 4 pounds per square foot per hour on a grate having 7 square feet of surface, was to secure a superheating of 134 degrees, attended by a reduction of 9.2 per cent. in the evaporative efficiency. In another experiment on the same plant, a superheating of 163 degrees was attended by a reduction of 10.2 per cent. in the evaporation per pound of total fuel used.

The same superheating pipes, placed over an independent furnace and surmounted by a heater made of small pipes, presenting an aggregate of 60 square feet of surface, were attached to a single 80 horse-power boiler of the horizontal tubular type, some 75 feet distant. A test on these, during which all the steam generated by the boiler passed through the superheater, showed that 473 pounds of anthracite coal were required to superheat 23,609 pounds of steam 228 degrees. The quantity of coal consumed in the boiler furnace, was 2657 pounds, and the rate of combustion 6.3 pounds per square foot of grate per hour. The rate of combustion in the superheater was 4.8 pounds per square foot of grate per hour. The evaporation of the boiler per pound of combustible from and at 212 degrees, was 10.95 pounds; temperature of feed water, 215 degrees; temperature of gases, 290 degrees; temperature of gases leaving superheater, 394 degrees. In this case a superheating of 228 degrees required the consumption of an additional 17.8 per cent. of coal in the independent furnace.

The quantity of coal required for superheating, in all these cases, is greater than that which would be expected if account be taken simply of the additional heat of the steam. The difference, however, is not more than can fairly be attributed to a slightly moist condition of the steam generated by the boiler and to loss of heat produced by radiation from the pipe leading from the boiler to the superheater, and by radiation from the superheater itself.

Boiler No. 2.

Kind of boiler,	Horizontal return tubular.
Number used,	Two.
Horse-power (collective, basis 12 sq. ft.),	One hundred and three.
Kind of coal,	Bituminous, Cumberland.
Age,	Several years.

Boiler No. 2 consists of two 48 inch boilers, set over a furnace common to both, in the manner shown in the following cuts. The draught of the chimney, which is deficient, is aided by a blower, which discharges the air beneath the grate. Except in these particulars, the boiler, as arranged for the first test, calls for no special comment. The dotted lines shown in the longitudinal section represent heating surfaces, which were introduced after the first test was made. These surfaces consisted, in part, of a row of pipes, connected together with return bends, placed behind the bridge wall, and, in part, of a cylinder, which was located in the place ordinarily occupied by the bridge wall at the back end of the grate. The total area of the surface thus introduced amounted to 60 square feet. The feed water on leaving the pump was first discharged into the pipes and bridge wall cylinder. After circulating through these, it was discharged into the steam space at the top of the boiler. At the time of the introduction of these pipes, the bridge wall at the rear end of the boiler was removed altogether, and the opening above the wall at the front was increased. A change was also made in the speed of the blower, which, on the first test, was at times too low for the desired draught.

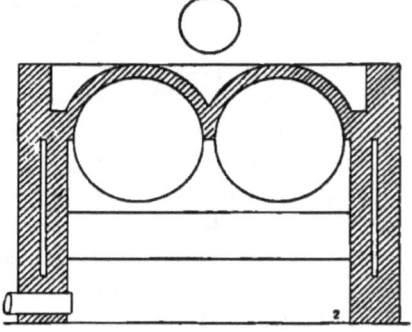

BOILER No. 2, CROSS SECTION THROUGH FURNACE.

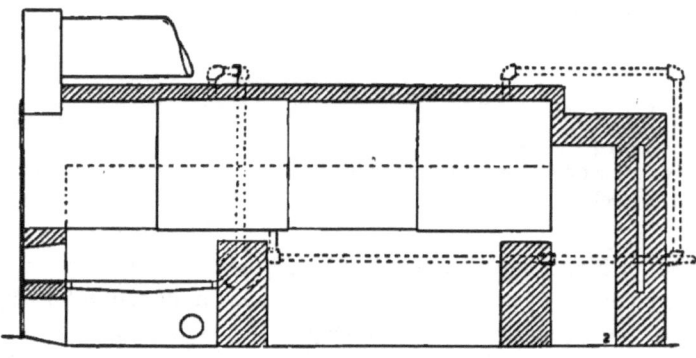

BOILER No. 2, LONGITUDINAL SECTION.

Dimensions of Boiler No. 2.

Diameter of each shell,	48 in.
Length between heads and length of tubes,	15 ft.
Number of tubes (collective) 3 inches outside diameter,	99
Area of heating surface,	1,235 sq. ft.
Area of grate surface (common to two boilers),	40.5 sq. ft.
Area through tubes,	4.1 sq. ft.
Area through flue,	2.7 sq. ft.
Chimney height,	130 ft.
Distance of grate to shell,	20 in.
Ratio of heating surface to grate,	30.5 to 1
Ratio of grate to tube area,	9.9 to 1
Ratio of grate to flue area,	15 to 1

Results of Tests. Boiler No. 2.

	Test No. 3.	Test No. 4.
Manner of start and stop,	Running.	Running.
Kind of run,	Continuous.	Continuous.
Duration, hrs.	5.25	6
Coal consumed, lbs.	2,302	2,441
Percentage of ash, per cent.	9.6	10.5
Water evaporated, lbs.	19,550	27,955
Coal per hour lbs.	438.5	406.8
Coal per hour per square foot of grate, lbs.	10.8	10.0
Water per hour, lbs.	3,723.8	3,659.2
Water per hour per square foot of heating surface, lbs.	3.01	2.96
Horse power developed, H. P.	117.3	115.1
Boiler pressure, lbs.	68	63
Temperature of feed-water, deg.	160	160
Temperature of escaping gases, deg.	355	346
Percentage of moisture in steam, per cent.	0.6	3.0
Water per pound of coal, lbs.	8.49	8.99
Water per pound of coal from and at 212 degrees, lbs.	9.22	9.77
Water per pound of combustible from and at 212 degrees, lbs.	10.19	10.91

The principal object of the tests on Boiler No. 2, was to show the effect produced by increasing the heating surface of the boiler in the manner noted. This object was not fully realized, because changes were made in the setting of the boilers, in addition to the change in the amount of heating surface. One result produced was to increase the percentage of moisture in the steam. This was probably due to the discharge of the feed water into the steam space. The increase amounted to 2.4 per cent. Making allowance for this difference in the quality of the steam, the evaporation was increased on the second test 3.5 per cent. Judging by the low temperature of the escaping gases, the boiler absorbed nearly the whole of the heat of the products of combustion, and in this respect the result obtained is all that could be desired. The water evaporated per pound of combustible is less than the best results obtained from boilers of the same type, using Cumberland coal. It is inferred from the high proportion of ash, that the coal was of an inferior quality, and that this fact had much to do with the low degree of economy.

Boiler No. 3.

Kind of boiler, ... { Horizontal return tubular.
Number used, ... Six.
Horse-power (collective, . { Three hundred and thirty-eight.
basis 12 sq. ft.),
Kind of coal, ... { Anthracite Lackawanna, broken.
Age, ... Six years.

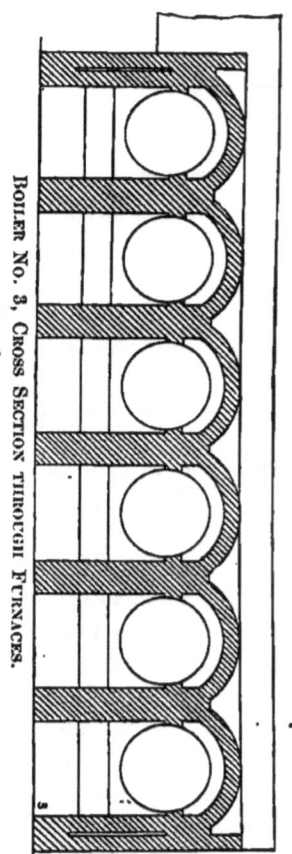

Boiler No. 3, Cross Section Through Furnaces.

Boiler No. 3 consists of a plant of six 48 inch boilers, set in one battery of brick work, as shown in the following cuts. The products of combustion on leaving the tubes pass over the top of the shells to the main flue at the rear, and this arrangement provides a small amount of heating surface exposed to the escaping gases. There is no provision for cleaning the tops of the shells, and this surface is doubtless covered by a deposit of soot, and thereby rendered inefficient.

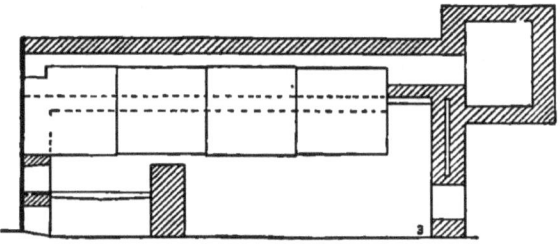

BOILER NO. 3, LONGITUDINAL SECTION.

Dimensions of Boiler No. 3.

Diameter of each shell,	48 in.
Length between heads and length of tubes,	15 ft.
Number of tubes (collective) 3 inches outside diameter,	288
Area of water-heating surface,	4,056 sq. ft.
Area of steam-heating surface,	540 sq. ft.
Area of grate surface,	120 sq. ft.
Area through tubes,	11.9 sq. ft.
Area through chimney,	11.1 sq. ft.
Height of chimney,	100 ft.
Ratio of water-heating surface to grate,	33.8 to 1
Ratio of steam-heating surface to grate,	4.5 to 1
Ratio of grate to tube area,	10.1 to 1

Results of Test. (Average of three.) Boiler No. 3.

	Test No. 5
Manner of start and stop and kind of run,	Ordinary.
Duration,	12 hrs.
Coal fired (including wood equivalent),	16,003 lbs.
Percentage of ash (including some unburned coal),	16.7
Water evaporated,	136,034 lbs.
Coal per hour,	1,333.6 lbs.
Coal per hour per square foot of grate,	11.11 lbs.
Water per hour,	11,336.1 lbs.
Water per hour per square foot of water-heating surface,	2.79 lbs.
Horse-power developed,	345
Boiler pressure,	84.6 lbs.
Temperature of feed-water,	198.6 deg.
Temperature of escaping gases,	482 deg.
Water per pound of coal,	8.50 lbs.
Water per pound of coal from and at 212 degrees,	8.96 lbs.
Water per pound of combustible from at 212 degrees,	10.73 lbs.

The test on Boiler No. 3 may be regarded as typical of the work done by a large plant of boilers, using anthracite coal, which had been in service a number of years. The ratio of water heating surface to grate surface, which is 33.8 to 1, does not show what would be called a deficiency of surface under proper conditions, but in this case the surface does not appear to have been of sufficient area to absorb the whole heat of the products of combustion. The temperature of the gases was 482 degrees, and this high figure, taken in connection with the fact of the boilers' age, is indicative of the presence of scale on the interior surfaces. The loss of heat thus produced is evidently the cause of the somewhat low economic result.

Boiler No. 4.

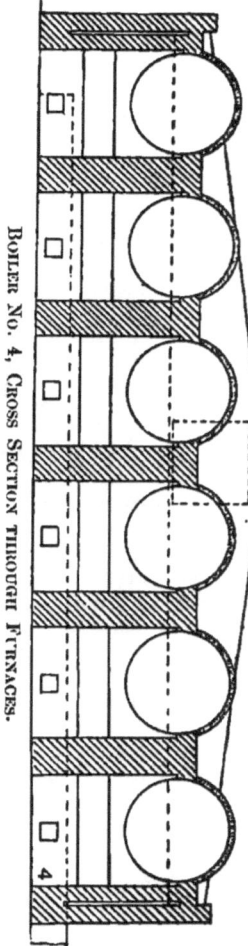

BOILER NO. 4, CROSS SECTION THROUGH FURNACES.

Kind of boiler, . Horizontal return tubular.
Number used, . . Six.
Horse-power (collective, basis 12 sq. ft.), } Four hundred and thirty.
Kind of coal, . . { Mixture of Anthracite Screenings and Nova Scotia Culm.
Age, ... Two years.

Boiler No. 4 embraces a plant of six horizontal tubular boilers, set in one battery of brick work, as shown in the following cuts. These boilers are arranged in such a manner that the products of combustion, instead of passing from the smoke arch at the front into the flue, in the manner usually followed, are carried back to the rear of the boiler through the two upper rows of tubes. The number of tubes in each boiler is 77. Those in the lower rows, which carry the products of combustion forward, number 51, and those in the upper rows, referred to, number 26. The main flue is located at the rear end and is made of wrought iron. This arrangement of tubes, and

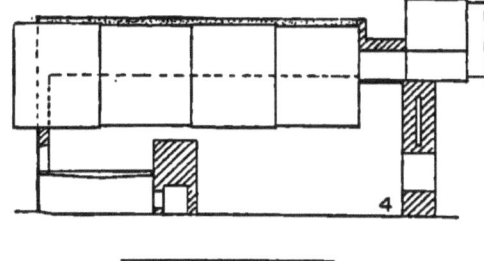

BOILER No. 4, LONGITUDINAL SECTION.

the use of a fine grade of coal, necessitated the employment of a blower to aid the draught, the air being discharged into the ash pits.

The main flue contains a coil of pipe having a total surface amounting to 300 square feet, through which the feed water was passed before entering the boiler.

Dimensions of Boiler No. 4.

Diameter of each shell,	60	in.
Length between heads and length of tubes,	15	ft.
Number of tubes (collective) 3 inches outside diameter,	462	
Area of heating surface,	6,204	sq. ft.
Area of grate surface,	150	sq. ft.
Area through tubes (at smallest section),	6.42	sq. ft.
Area through chimney,	16	sq. ft.
Height of chimney,	120	ft.
Ratio of heating surface to grate surface,	41.4	to 1
Ratio of grate surface to smallest tube area,	23.4	to 1
Ratio of grate surface to chimney area,	9.2	to 1

Results of Tests. Boiler No. 4.

		Test No. 6.
Manner of start and stop and kind of run,		Ordinary.
Duration,	10.5	hrs.
Coal consumed, moist (including wood equivalent),	11,265	lbs.
Proportion of bituminous coal to whole,	.178	
Percentage of ash and waste coal,	15.4	
Water evaporated,	78,849	lbs.
Coal per hour,	1,072.8	lbs.
Coal per hour per square foot of grate,	7.15	lbs.
Water per hour,	7,509.4	lbs.
Water per hour per square foot of heating surface,	1.21	lbs.
Horse-power developed,	250.3	lbs.
Boiler pressure,	45	lbs.
Temperature of feed-water,	89	deg.
Temperature of escaping gases,	305	deg.
Draught suction of chimney,	3-8	in.
Draught pressure of fan,	0 to 1-2	in.
Water per pound of coal,	7.00	lbs.

Water per pound of coal from and at 212 degrees, . . 8.17 lbs.
Water per pound of combustible from and at 212 degrees, . 9.66 lbs.

The interest in the test of Boiler No. 4, centers in the arrangement of tubes noted, the kind of fuel burned, and the use of the blower for increasing the draught. The only effect of the reduced opening through the tubes, which is noticeable, is the small amount of power developed, this being but little over one-half of the rated power of the boilers. The fuel contained a large proportion of anthracite screenings, which produced a large percentage of ash. The heat was well absorbed by the boilers, the escaping gases passing to the chimney at about the temperature of the steam. The evaporative result, considering the grade of fuel used, and taking into account the fact that the fuel contained about 4 per cent. of moisture, compares favorably with that obtained with the best grades of coal.

The heat absorbed by the water in passing through the coil in the flue, served to increase its temperature 13 degrees. The water was fed by an injector, the temperature of its discharge, or approximately that of the water entering the coil, being 155 degrees.

Subsequent tests on this boiler, made at intervals of about a year, and using different kinds of cheap-grade anthracite coal, gave the following results:

Kind of anthracite coal,	Screenings.	Buckwheat.	Pea and Dust.
Proportion of Nova Scotia coal, ...	.14	.055	.105
Percentage of ash, per cent.	20	18	19.8
Horse-power developed, H. P.	380	368	561
Temperature of escaping gases, ... deg.	348	337	407
Water per pound of coal from and at 212 degrees, . lbs.	7.95	8.92	8.26
Water per pound of combustible from and at 212 degrees, ...	9.93	10.87	10.32
Draught pressure at blower, in.	–	–	0 to 1¼

The amount of power developed by the boilers on the supplementary tests was increased. On the last test the power exceeded that of the rated horse power, and it may be noted that this was about the maximum which, under the unfavorable circumstances noted, could be obtained. The increased amount of power obtained was accompanied by an increase in the amount of ash, due probably to a greater amount of coal wasted by dropping through the grates. It is seen that as the power increases, the temperature of the escaping gases becomes higher, reaching a maximum in the last test of 407 degrees. This is not a high figure for the amount of power developed, and it indicates that the special arrangement of tubes secured a more thorough absorption of the heat than would otherwise occur.

Boiler No. 5.

Kind of boiler,	Horizontal return tubular.
Number used,	One.
Horse-power, (basis 12 square feet),	Fifty-four.
Age,	Several years.

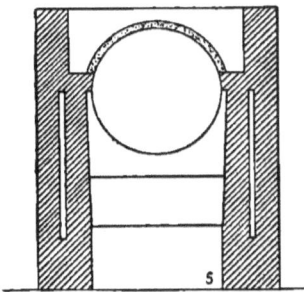

BOILER NO. 5, CROSS SECTION THROUGH FURNACES.

Boiler No. 5 is a horizontal return tubular boiler, set in brick-work in the manner shown in the following cuts. It is the end boiler of a plant of the same kind, and the boiler next to it was in operation during the progress of the tests. The boiler is an old one, and the heating surfaces are somewhat covered with scale.

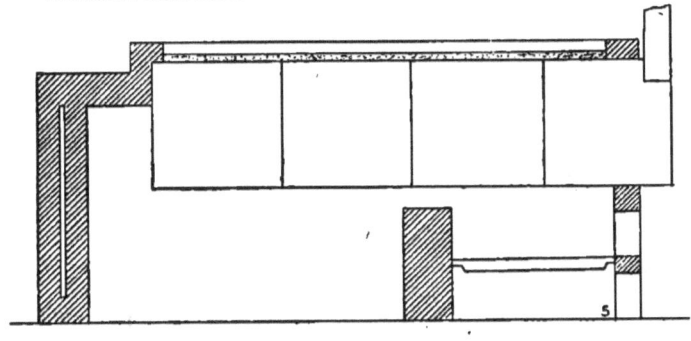

BOILER NO. 5, LONGITUDINAL SECTION.

BOILER No. 5.

Dimensions of Boiler No. 5.

Diameter of each shell,	48 in.
Length between heads and length of tubes,	15 ft.
Number of tubes three inches outside diameter,	49
Area of heating surface,	644 sq. ft.
Area of grate surface,	20 sq. ft.
Area through tubes,	2 sq. ft.
Width of air spaces and metal bars in grates,	3-8 in.
Distance of grate to shell,	27 in.
Distance of bridge wall to shell,	8 in.
Ratio of heating surface to grate surface,	32.2 to 1
Ratio of grate surface to tube area,	9.9 to 1

Results of Tests. Boiler No. 5.

	Test No. 7.	Test No. 8.	Test No. 9.	Test No. 10.
Kind of coal	George's Creek, Cumberland, Bituminous.	Delaware and Hudson, Lackawanna, Anthracite, Broken.	Mixture 1 part Cumberland, 2 parts Pea and Dust.	Mixture 1 part Nova Scotia Culm, 2 parts Pea and Dust.
Manner of start and stop and kind of run,	Ordinary.	Ordinary.	Ordinary.	Ordinary.
Duration, hrs.	11.2	11.2	11.2	11.2
Coal consumed, dry (including wood equivalent), lbs.	2,277	2,266	1,878	2,080
Percentage of ash, per cent.	11.1	14.2	21.3	26.2
Water evaporated, lbs.	20,464	18,385	13,144	13,230
Coal per hour, lbs.	202.4	201.4	167	184.9
Coal per hour per square foot of grate, lbs.	10.1	10.1	8.3	9.2
Water per hour, lbs.	1,819	1,634.3	1,168.4	1.176.1
Water per hour per square foot of heating surface, lbs.	2.8	2.5	1.8	1.8
Horse-power developed, H. P.	60	53.9	38.5	38.8
Boiler pressure, lbs.	69.4	69.4	71.7	71.6
Temperature of feed-water, deg.	119.4	120.5	119.7	120.4
Temperature of escaping gases, deg.	435	443	430	406
Number of firings,	17	12	14	15
Number of times slicing bar was used,	15	4	8	- 18
Water per pound of coal, lbs.	9.08	8.20	7.10	6.46
Water per pound of coal from and at 212 degrees, lbs.	10.25	9.24	7.99	7.27
Water per pound of combustible from and at 212 degrees, lbs.	11.52	10.76	10.18	9.87

NOTE.—The mixed fuels were wet before firing with 5 per cent. of their weight of water.

The tests on Boiler No. 5 were made to determine the relative economy between several different kinds of coal. The Cumberland coal gave the highest evaporative result and the mixed coal the lowest. The evaporation per pound of Cumberland coal was 10 per cent. higher than that with Anthracite coal, 28 per cent. higher than that with the mixture of pea and dust and Cumberland, and 41 per cent. higher than that with the mixture of pea and dust and Nova Scotia culm. Judging by the percentages of ash, which are all comparatively large, the coal was in each case of somewhat inferior quality. This is especially marked in the Cumberland coal, which gave 11.1 per cent. of ash, and in both of the mixed fuels. The unfavorable effect produced by the coating of scale on the heating surfaces is seen in the comparatively high temperature of the escaping gases.

An important feature of the results is the relative amount of power developed by the various fuels. The dampers were kept wide open during each test and the boilers produced a maximum quantity of steam under each condition. The Cumberland coal produced the largest amount, viz., 60 horsepower, and this is 22 per cent above the nominal capacity. Then comes the anthracite coal and finally the mixed fuels. The mixed fuels produced 22 per cent below the rated power of the boilers. Another feature is the labor involved in using the various fuels, as indicated by the number of times the slicing-bar was employed in breaking up the bed of coal. The anthracite coal gives the most favorable showing, while the mixture of culm and pea and dust gives the least favorable showing. The difference in the labor involved in slicing a fire four times per day and eighteen times per day is considerable.

The actual economy obtained with the different fuels is shown by the following table, which gives the cost of coal required to generate 30,000 pounds of steam, according to quotations of prices which ruled at the time of the tests.

	Cumberland.	Anthracite.	Mixture Cumberland and Pea and Dust.	Mixture Culm and Pea and Dust.
Cost of coal per ton of 2,240 pounds, . .	$5.90	$5.60	$4.47	$3.75
Cost of coal required to evaporate 30,000 pounds of water from and at 212 degrees, . . .	7.70	8.11	7.50	6.88

The extreme difference in these figures is $1.23, which corresponds to 15 per cent., and this occurs between the anthracite coal, which is the most easily worked, and the mixed fuel containing culm, which requires the most labor in firing. If one fireman be assumed capable of handling ten tons of anthracite coal, broken size, per day, and his wages are $2.00 per day, the cost of fireman's labor represents 3.6 per cent. of the cost of fuel. A saving of 15 per cent. in the cost of fuel, when using 10 tons per day of the mixture named, is sufficient to cover the increased cost of labor due to the employment of another fireman, and still have 11.4 per cent. remaining for a net saving.

Additional tests were made on Boiler No. 5, using Franklin and Kalmia coal, each of which was quoted at $6.75 per ton. The Franklin coal gave an evaporation of 9.78 pounds of water from and at 212 degrees per pound of coal, and the Kalmia coal, 9.68 pounds. The percentage of ash in each of these coals was 10.8 per cent., while in the Lackawanna, on Test No. 8 the percentage was 14.2. It is evident that the improved performance with these coals was due simply to the greater proportion of combustible matter which the coal contained.

Boiler No. 6.

Kind of boiler, ... Horizontal return tubular.
Number used, ... One.
Horse-power (basis 12 square feet), . Sixty-three.
Kind of coal, . . . { Mixture two-parts Screenings, one part Nova Scotia Culm.
Age, . . . Several years.

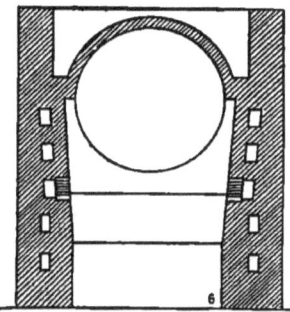

BOILER No. 6, CROSS SECTION THROUGH FURNACE.

Boiler No. 6 is a horizontal tubular boiler, set in brick-work in the manner shown in the following cuts. It is located at the end of a battery of boilers of the same kind, and during the progress of the tests all the boilers were in operation. The arrangement of the setting is such that air is admitted over the fire in a finely divided state. For this purpose perforated plates are introduced in the side walls of the furnace and in the top of the bridge wall, and the air is supplied through these plates from ducts extending back and forth through the walls.

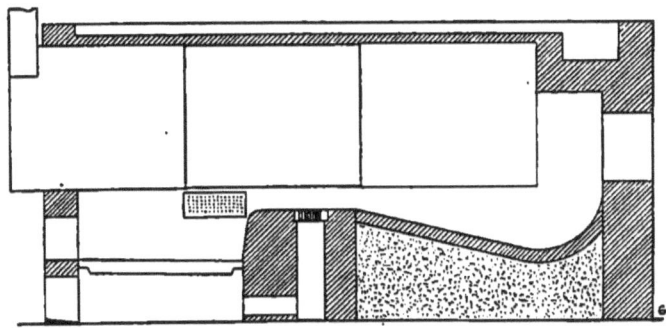

BOILER No. 6, LONGITUDINAL SECTION.

Dimensions of Boiler No. 6.

Diameter of shell,	54	in.
Length between heads and length of tubes,	15	ft.
Number of tubes 3 inches outside diameter,	60	
Area of heating surface,	757	sq. ft.
Area of grate surface,	22.5	sq. ft.
Area through tubes,	2.5	sq. ft.
Width of air spaces and metal bars in grates,	3-8	in.
Distance of grate to shell,	27	in.
Distance of bridge wall to shell,	8	in.
Ratio of heating surface to grate surface,	33.7	to 1
Ratio of grate surface to tube area,	9	to 1

BOILER No. 6.

Results of Tests. (Average of two days.) *Boiler No. 6.*

	Test No. 11.	Test No. 12.
Conditions regarding air passages,	Air passages open.	Air passages closed.
Manner of start and stop and kind of run,	Ordinary.	Ordinary.
Duration, ... hrs.	11.4	10.8
Coal consumed, dry (including wood equivalent), ... lbs.	2,820	2,673
Percentage of ash, ... per cent.	14.3	16.1
Water evaporated, ... lbs.	17,828	16,557
Coal per hour, ... lbs.	297.9	248.7
Coal per hour per square foot of grate, . lbs.	11.02	11 05
Water per hour, ... lbs.	1,567.7	1,541.3
Water per hour per square foot of heating surface, ... lbs.	2.06	2.03
Horse-power developed, . . . H. P.	53.7	52.8
Boiler pressure, ... lbs.	68.9	68.7
Temperature of feed-water, . . . deg.	72.2	71.6
Temperature of escaping gases, . . deg.	453	461
Draught suction, ... in.	0.36	0.36
Water per pound of coal, ... lbs.	6.32	6.19
Water per pound of coal from and at 212 degrees, ... lbs.	7.44	7.30
Water per pound of combustible from and at 212 degrees, ... lbs.	8.69	8.69

NOTE.— The coal when fired contained 5 per cent. of its weight of water.

The object of the tests on Boiler No. 6 was to determine the effect of excluding the air from the ducts in the walls, when burning a mixture of anthracite screenings and Nova Scotia culm. One test was made with the passages open in the ordinary manner, and another with the entrance to the ducts closed. During the first test the air registers, with which the fire doors were provided, were closed, and during the second test the registers were open a part of the time. The comparative tests are thus made between two systems of introducing air above the fuel, one the special method for which the boiler was set, and the other the common method of introducing it through the fire door. The evaporative results of these tests, which are an average of two days' run in each case, do not show a material difference. Basing the results on the evaporation per pound of coal, there appears to be a loss of about 2 per cent in closing the air ducts and using the air registers in the fire door. Basing the results on the evaporation per pound of combustible, there is an exact agreement between the two.

The results of this comparison show that the manner in which the gaseous products of the furnace appear to be consumed, as viewed by the eye of an observer, does not necessarily indicate much as to the true economy with which the combustion takes place. As thus viewed there was a marked improvement in the character of the combustion when the air passages were open, over the appearance when they were closed.

The boilers developed less than their rated capacity, even with a constant draught of 3-8 of an inch water pressure, which is the full draught of a 100 ft. chimney. The evaporation per pound of coal in both tests was much below that obtained with the best grades of coal. This is not surprising in view of the nature of the fuel, which is liable to be of inferior quality.

Boiler No. 7.

Kind of boiler,	Horizontal return tubular.
Number used,	Six.
Horse-power (collective, basis 12 sq. ft.)	Four hundred and seventy.
Kind of coal,	Anthracite Screenings 3 parts, Cumberland bituminous 1 part.
Age,	Five months.

Boiler No. 7 embraces a plant of six horizontal return tubular boilers, set in one battery of brick work. The style of setting is that shown in the cut of Boiler No. 6. Air is supplied above the fuel through perforations in the side walls and bridge wall, in the same manner as in that boiler. The air ducts in this case are supplied through sheet iron pipes, which are placed in the flue. These present about 3 per cent. as much surface to the heat of the gases as the area of the heating surface in the boilers. The air supplied above the fuel is thus made to utilize some of the heat which would otherwise be wasted.

BOILER No. 7.

Dimensions of Boiler No. 7.

Diameter of each shell,	60	in.
Length between heads and length of tubes,	15	ft.
Number of tubes (collective) 3 1-2 inches outside diameter,	348	
Area of heating surface,	5,046	sq. ft.
Area of grate surface,	150	sq. ft.
Area through tubes,	20	sq. ft.
Height of chimney,	150	ft.
Width of air spaces and metal bars in grates,	3-8	in.
Ratio of heating surface to grate surface,	37.6	to 1
Ratio of grate surface to tube area,	7.5	to 1

Results of Tests. (Two days.) Boiler No. 7.

		Test No. 13.
Manner of start and stop and kind of run,		Ordinary.
Duration,	13.5	hrs.
Coal consumed, dry (including wood equivalent),	20,476	lbs.
Percentage of ash,	14.2	per cent.
Water evaporated,	163,705	lbs.
Coal per hour,	1,516.7	lbs.
Coal per hour per square foot of grate,	10.1	lbs.
Water per hour,	12,125.9	lbs.
Water per hour per square foot of heating surface,	2.1	lbs.
Horse-power developed,	404.2	H. P.
Boiler pressure,	86	lbs.
Temperature of feed-water,	102	deg.
Temperature of escaping gases,	410	deg.
Water per pound of coal,	7.99	lbs.
Water per pound of coal from and at 212 degrees,	9.18	lbs.
Water per pound of combustible from and at 212 degrees,	10.70	lbs.

NOTE. — The coal when fired contained 5 per cent. of water.

The results of the test on Boiler No. 7 furnish an example of the performance of a low grade of fuel in a large plant of new boilers, where the conditions were favorable for obtaining a good result. The evaporation per pound of coal compares favorably with that produced in many cases with the best grades of anthracite coal, and this is all that could be desired from fuel composed largely of screenings. In this case the screenings were evidently of good quality.

Boiler No. 8.

Kind of boiler,	Horizontal tubular.
Number used,	One.
Horse power (basis 12 square feet),	One hundred.
Kind of coal,	Anthracite pea and dust 3 parts, George's Creek Cumberland 1 part.
Age,	New.

Boiler No. 8 is a horizontal tubular boiler, provided with a detached furnace, and the general features of the boiler and setting are shown in longitudinal section in the following cut. The furnace has the same location with respect to the tubes as that found in the locomotive type of boiler. The products of combustion pass forward through the tubes, and after returning beneath the shell, they enter the chimney from the front end. The boiler is provided with forced draught, which is supplied by a blower discharging air under the grates. The air from the blower first passes through a pipe in the flue space beneath the shell, which presents a heating surface amounting to 15 per cent of the area of that in the boiler. The temperature of the air supplied under the grates when the blower was in operation, was increased by this means about 40 degrees.

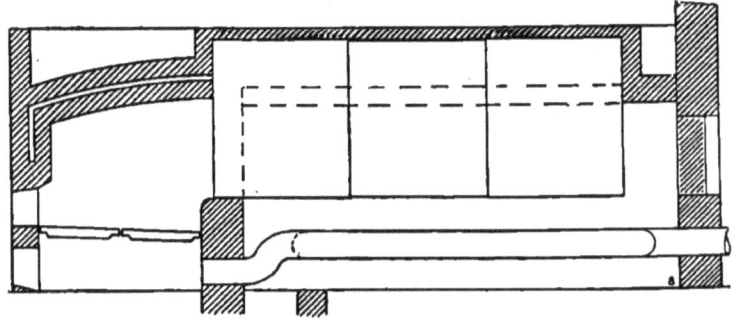

BOILER NO. 8, LONGITUDINAL SECTION.

BOILER No. 8.

Dimensions of Boiler No. 8.

Diameter of shell,	72 in.
Length between heads and length of tubes,	15 ft.
Number of tubes 4 inches outside diameter,	70
Area of heating surface,	1,189 sq. ft.
Area of grate surface,	36 sq. ft.
Area through tubes,	5.3 sq. ft.
Area through chimney,	5.9 sq. ft.
Height of chimney,	60 ft.
Width of air spaces and metal bars in grates,	Air 7-16 in., metal 3-8 in.
Ratio of heating surface to grate surface,	33 to 1
Ratio of grate surface to tube area,	6.7 to 1

Results of Tests. Boiler No. 8.

	Test No. 14.	Test No. 15.
Conditions as to draught,	Natural draft.	Forced draft.
Manner of start and stop,	Ordinary with preliminary heating.	Ordinary with preliminary heating.
Kind of run,	Continuous.	Continuous.
Duration, ... hrs.	7.2	7.5
Coal consumed, dry (including wood equivalent), ... lbs.	2,750	3,430
Percentage of ash, . . . per cent.	11	9.6
Water evaporated, ... lbs.	21,876	27,869
Coal per hour, ... lbs.	379.4	457.4
Coal per hour per square foot of grate, lbs.	10.5	12.7
Water per hour, ... lbs.	3,017.4	3,715.9
Water per hour per square foot of heating surface, ... lbs.	2.5	3.1
Horse-power developed, . . H. P.	105.6	130
Boiler pressure, ... lbs.	70	71
Temperature of feed-water, . . deg.	51	52
Temperature of escaping gases, . deg.	395	400
Water per pound of coal, . . lbs.	7.95	8.12
Water per pound of coal from and at 212 degrees, ... lbs.	9.54	9.74
Water per pound of combustible from and at 212 degrees, ... lbs.	10.73	10.79

NOTE. — The coal was wet before firing, and the weight was increased by this means about 6 per cent.

The tests on Boiler No. 8 were made to determine the comparative economy of forced draught with heated air, and natural draught, where the fuel was largely composed of anthracite screenings, together with the general result produced by the use of a detached furnace. An exact comparison cannot be made between the two tests, owing to the different conditions in respect to capacity, 130 horse-power being developed with forced draught, and 105.6 horse-power with natural draught. Taking the results, however, as they stand, there is a gain amounting to 2 per cent. in favor of forced draught, figured on coal, and 0.5 per cent. figured on combustible. The small advantage produced by the forced draught may be attributed wholly to the heated air which was used. It is to be borne in mind that with forced draught there is a chance for loss from excessive supply of air at times when the bed of coal is allowed to burn through to the grate. The loss from this source is liable to be greater with forced draught than with natural draught, on account of the greater quantity and force of the air supplied by the blower. Little, if any benefit, for this reason, can be expected under ordinary conditions of firing from a forced draught, and the results of the tests bear out these expectations.

The evaporative results, taken by themselves, show a favorable performance of the boiler, for the class of fuel that was used. Compared with the results obtained from boilers set in the ordinary way, with the fire box beneath the shell, the figures do not show any special improvement, and the conclusion may be drawn that the use of a detached furnace is not especially advantageous. In this form of furnace the fire is surrounded by highly heated brick work, which keeps the furnace at a high temperature. In the ordinary setting the temperature is much reduced by the close proximity of the heating surfaces of the boiler to the furnace. The use of a detached furnace, on account of the higher temperature, would be expected to secure the best result. The same cause, however, which produces the high temperature, leads to a considerable amount of loss of heat, on account of radiation from the top and sides of the furnace, and this stands in the way of realizing the whole benefit, which would otherwise be expected from this system.

Another test was made on Boiler No. 8, using screenings alone with forced draught (the screenings being wet 7 per cent. before firing). The percentage of ash was 15 per cent.; the horse-power developed was the same as in test No. 14; the temperature of the escaping gases was 356 degrees (39 degrees lower than that of test No. 14). The evaporation from and at 212 degrees was 8.17 pounds per pound of coal and 9.63 pounds per pound of combustible. It is not certain that the screenings here used was of the same quality (though of the same name) as those used in the mixture tests; nevertheless, a large gain appears to have been produced by the introduction of 25 per cent. of Cumberland coal. The increase, figured upon combustible, is about 11 per cent.

Boiler No. 9.

Kind of boiler,	Horizontal return tubular.
Number used,	One.
Horse-power (basis 12 square feet),	Seventy-four.
Age,	Two months.

Boiler No. 9 is of the horizontal tubular type, though unlike the ordinary form in being provided with a water leg, which forms the front wall of the furnace. The arrangement of the water leg, and the general features of the boiler, together with the manner in which it is set, are shown in the following cuts. The water leg extends a short distance below the level of the grates, and at the proper point it is provided with an opening through which the coal is fired. The setting is so arranged, as shown, that air is supplied to the products of combustion as they pass the bridge wall. The air is introduced through a pipe laid inside the brick work, and distributed to cast iron globes, which rest upon the top of the bridge wall. The surfaces of these globes are perforated. A jet of steam is introduced into the end of the supply pipe, and a mixed current of air and steam is thereby distributed through the burning gases. The size of the steam pipe is 3-8 of an inch, and its outlet is drawn down to a

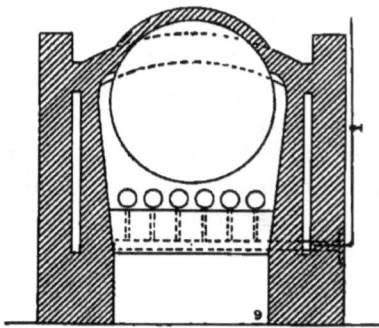

BOILER NO. 9, CROSS SECTION THROUGH FURNACE.

diameter of about 1-4 of an inch. The air which is fed to the bridge wall pipe first passes through the air spaces in the side walls of the setting, by which means it is somewhat heated.

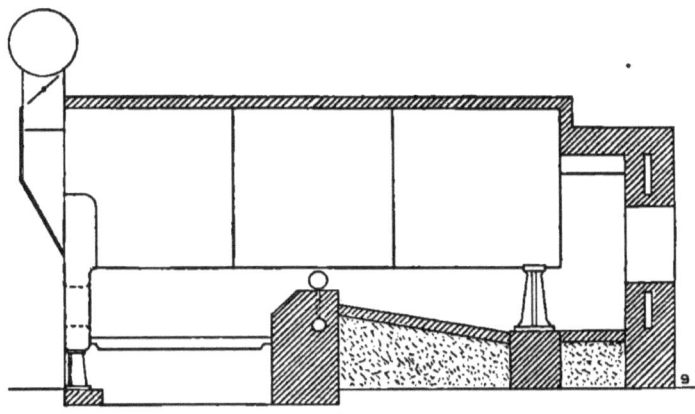

BOILER No. 9, LONGITUDINAL SECTION.

Dimensions of Boiler No. 9.

Diameter of each shell,	60 in.
Length between heads and length of tubes,	15 ft.
Number of tubes 3 inches outside diameter,	66
Area of heating surface,	890 sq. ft.
Area of grate surface,	25.7 sq. ft.
Area through tubes,	2.7 sq. ft.
Area through chimney,	7 sq. ft.
Height of chimney,	90 ft.
Width of air spaces and metal bars in grate,	Air 3-8 in., metal 1-2 in.
Distance of grate to shell,	26 in.
Distance of flat bridge wall to shell,	9 in.
Ratio of heating surface to grate surface,	34.6 to 1
Ratio of grate surface to tube area,	9.4 to 1

Results of Tests. Boiler No. 9.

	Test No. 16.	Test No. 17.	Test No. 18.
Kind of coal,	George's Creek, Cumberland, Bituminous.	Lehigh Egg, Anthracite.	Equal parts Anthracite, Screenings, & Cumberland.
Manner of start and stop and kind of run,	Ordinary.	Ordinary.	Ordinary.
Duration, ... hrs.	14	17.5	14.5
Coal consumed, dry (including wood equivalent), . lbs.	6,559	6,309	5,102
Percentage of ash, . per cent.	6.6	9.4	13.5
Water evaporated, . . lbs.	56,555	53,153	39,658
Coal per hour, . . . lbs.	498.5	360.5	351.5
Coal per hour per square foot of grate, ... lbs.	18.2	14	13.7
Water per hour, . . . lbs.	4,039.7	3,037.3	2,735.1
Water per hour per square foot of heating surface, . lbs.	4.5	3.4	3.1
Horse-power developed, H. P.	143.8	105.5	95.1
Boiler pressure, . . . lbs.	66	62	65
Temperature of feed-water, deg.	40	40	42
Temperature of escaping gases, deg.	40.3	349	343
Percentage of moisture in steam, per cent.	0.3	–	–
Water per pound of coal, . lbs.	8.62	8.42	7.77
Water per pound of coal from and at 212 degrees, . lbs.	10.43	10.18	9.39
Water per pound of combustible from and at 212 degrees, lbs.	11.17	11.24	10.85

The tests on Boiler No. 9 were made to determine the relative economy and capacity of this type of boiler with different kinds of fuel. The results of the tests are chiefly interesting in showing the quantity of power which can be obtained from a boiler of the horizontal return tubular type under the different conditions of the tests. The chimney gave a draught corresponding to three-eights of an inch water pressure, and the damper was carried wide open on all the tests, so as to secure its full capacity. Moreover, the fires were well tended so that they should do their maximum work, being frequently broken up and sliced, not only when the bituminous coals were burned, but also when anthracite coal was burned. Under these conditions the developed power, when Cumberland coal was used, was nearly two times the rated power; when anthracite Lehigh was used it was nearly 50 per cent. larger, and when the mixture of equal parts of screenings and Cumberland coal was burned, it was nearly one-third larger. Compared among themselves the Cumberland coal developed 36 per cent. more power than the anthracite coal and 50 per cent. more power than the mixture.

The relative economy of these fuels is not fairly shown because of the large difference in the conditions as to capacity, especially in the case of the test of the Cumberland coal. The low result produced here is no doubt due to the loss attending the large capacity, this being shown in the comparatively high temperature of the escaping gases. Considering the large amount of power developed in the case of the test with anthracite coal, the result in this case shows a high degree of economy. The high rate of combustion, viz., 14 pounds per square foot of grate per hour, and the low temperature of the escaping gases, which was 349 degrees, are the two vital elements which contributed to this result.

A subsequent test was made to determine the effect produced upon the economy of the boiler by dispensing with the admission of air at the bridge wall. The test was made when using Cumberland coal, and the principal results obtained were as follows:—

Coal per hour per square foot of grate,	18.8	lbs.
Percentage of ash,	7	per cent.
Horse-power developed,	135	H. P.
Temperature of feed-water,	40	deg.
Temperature of escaping gases,	486	deg.
Water per pound of coal,	7.94	lbs.
Water per pound of coal from and at 212 degrees,	9.62	lbs.
Water per pound of combustible from and at 212 degrees,	10.34	lbs.

Comparing the results of this test with those of test No. 16, made with air in use, it appears that the evaporation per pound of coal was reduced nearly 8 per cent.

Boiler No. 10.

Kind of boiler,	Horizontal return tubular.
Number used,	One.
Horse-power (basis 12 square feet),	Fifty-five.
Kind of coal,	Anthracite Lehigh, broken.
Age,	Several years.

BOILER No. 10.

Boiler No. 10 is of the ordinary horizontal tubular type. The manner in which it is set is shown in the following cut, which is a longitudinal section. The top of the bridge wall is curved upward, so as to conform to the curve of the shell, and the space behind is filled and curved at the top in a similar manner. The products of combustion on leaving the smoke arch in front, pass through the flue space, over the top of the shell, and thence to the chimney. The manner in which this is arranged is similar to that shown in the cross section given in connection with Boiler No. 3. This arrangement secures a small amount of steam heating surface, though the position of this surface is such as to be more or less inefficient, on account of deposits of soot.

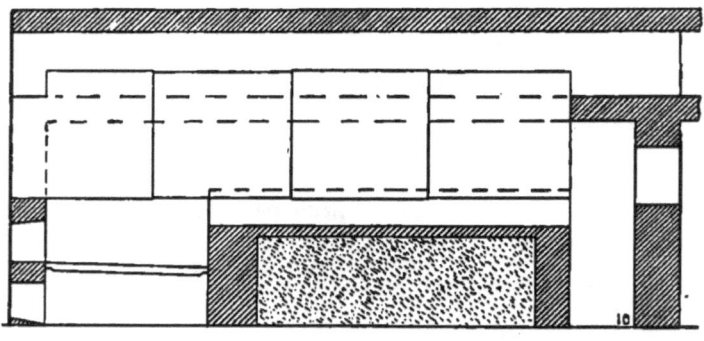

BOILER No. 10, LONGITUDINAL SECTION.

Dimensions of Boiler No. 10.

Diameter of each shell,	48 in.
Length between heads and length of tubes,	16 ft.
Number of tubes 3 inches outside diameter,	48
Area of water-heating surface,	656 sq. ft.
Area of steam-heating surface,	80 sq. ft.
Area of grate surface,	19.5 sq. ft.
Area through tubes,	2 sq. ft.
Width of air spaces and metal bars in grates,	3-8 in.
Distance of grate to shell,	24 in.
Distance of curved bridge wall to shell,	8 in.
Ratio of water-heating surface to grate surface,	33.5 to 1
Ratio of steam-heating surface to grate surface,	4.1 to 1
Ratio of grate to tube area,	9.7 to 1

Results of Tests, Boiler No. 10.

	Test No. 19.	Test No. 20.	Test No. 21.	Test No. 22.
Conditions,	Hand regulation of draught.	Automatic regulation of draught.	Grates with 50 per cent. air space.	Grates with 60 per cent. air space.
Manner of start and stop,	Ordinary.	Ordinary.	Ordinary.	Ordinary.
Kind of run,	Continuous.	Continuous.	Continuous.	Continuous.
Duration, . . . hrs.	10.2	10	9.5	9.7
Coal consumed (including wood equivalent,) lbs.	937	980	,623	1,643
Percentage of ash, per cent.	13.1	14.1	12 5	13.9
Water evaporated, . lbs.	8,638	9,263	15,935	15,597
Coal per hour, . . lbs.	91.9	91.3	170.8	168.5
Coal per hour per square foot of grate, . lbs.	4.7	4.7	8 7	8.6
Water per hour, . . lbs.	846.9	926.3	1,677.3	1,607.9
Water per hour per square foot of water-heating surface, . . lbs.	1.3	1.4	2.3	2.3
Horse-power developed, H.P.	25.1	27.8	50.3	48.2
Boiler pressure, . . lbs.	46	47	47	46
Temperature of feed-water, deg.	200	200	203	200
Temperature of escaping gases, . . . deg.	297	299	340	348
Water per pound of coal, lbs.	9.22	9.45	9.83	9.49
Water per pound of coal from and at 212 degrees, lbs.	9.57	9.81	10.17	9.85
Water per pound of combustible from and at 212 degrees, . . lbs.	11.01	11.42	11 63	11.44

Tests No. 19 and No. 20, made on Boiler No. 10, had for an object the determination of the economy of automatic regulation of the draught over hand regulation. On the test with hand regulation the method pursued was to alternately open and shut the damper through its full range during the whole progress of the test. When the damper was open it was kept in this position till the steam pressure rose to a certain figure determined upon beforehand. It was then closed and kept in that position till the pressure fell to a certain lower figure. Test No. 19 gives the average of three trials, on the first of which the variation in pressure, in the manner noted, amounted to 4 pounds, on the second, 7 pounds and on the third 1 pound. On the test with automatic regulation the type of regulator employed was that in which the pressure acts through a diaphragm and a system of levers upon the damper. The extreme variation of pressure was limited to 2 pounds. A comparison of the results of these tests shows that the automatic regulation secured an improvement over hand regulation amounting to 2.5 per cent. based on coal, and 3.7 per cent. based on combustible. It is a difficult matter to show by means of a test the working economy produced by automatic regulation, because so much depends upon the personal element when the draught is controlled by hand. On such a test, the fireman may have an incentive to do his best, which he might not have in ordinary work. For this reason the tests reveal the tendency which automatic regulation has to secure economy, rather than the actual gain which may be secured under working conditions. In this connection it may be added that tests were made on two large plants of boilers, to determine the economy of automatic regulation over hand regulation, and in one case the result was favorable to automatic regulation, while in the other case it was favorable to hand regulation.

The object of Tests No. 21 and No. 22 was to determine the effect which the type of grates may have upon the economy. The principal difference in the two grates lay in the proportion of air space. Both tests were made with automatic regulation of the draught. The grates with 50 per cent. air space gave 3.2 per cent. better result based on coal, and 1.7 per cent. better result based on combustible, than the grates with 60 per cent. air space. Subsequent tests were made under similar conditions, with a slower rate of combustion. In this case the gain due to the smaller air space was more marked, being about 8 per cent. From this showing it may be concluded that the slower the rate of combustion, the smaller should be the opening for draught through the grates. The gain probably comes about by preventing the introduction of too great an excess of air over that required for combustion.

Comparing Test No. 21 with Test No. 20, both of which were made with the same grates, it appears that the more rapid combustion on Test No. 21 secured a noticeable advantage. It amounted to 3.6 per cent. based on coal, and 1.9 per cent. based on combustible. Taking into account the slow rate of combustion on Test No. 20, which was 4.7 pounds per square foot of grate per hour, the performance of the boiler is excellent. This is significant when the small diameter of the shell, and the age of the boiler are taken into account.

Boiler No. 11.

Kind of boiler,	Horizontal return tubular.
Number used,	One.
Horse-power (basis 12 square feet),	Fifty-three.
Age,	Several years.

Boiler No. 11 is of the ordinary horizontal return tubular type. The arrangement of its setting is in all essential particulars like that shown in the cuts of Boiler No. 5, with the exception that this boiler is provided with a flush front. For the purpose of burning oil, as was done on Test No. 24, the furnace was arranged as shown in longitudinal section in the following cut. The oil burner consisted of a jet, having two openings, one of which was supplied with oil, and the other with steam. It was placed in the mouth of the furnace, and the oil was brought to it by force of gravity. The grates were covered with fire brick, and at the rear end of the furnace was placed a mass of loose bricks, upon which the flames of the burner were directed. The steam supplied to the burner was superheated by passing it through pipes placed on the bottom of the furnace, beneath the loose bricks referred to. The air for combustion entered through the doorways of the furnace, around the pipes leading to the burner, and the space surrounding the pipes was left open.

BOILER No. 11.

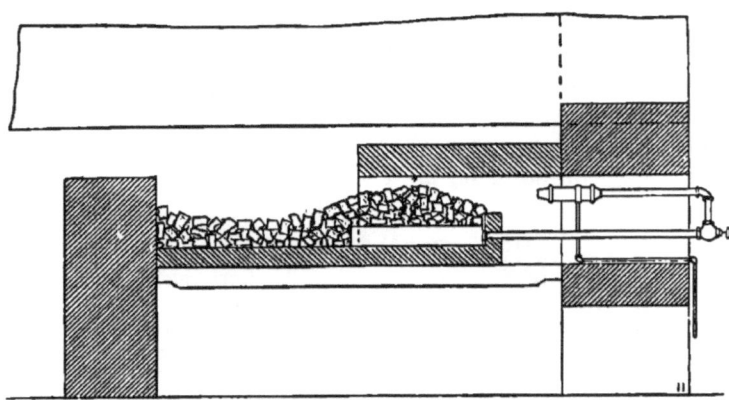

BOILER NO. 11, LONGITUDINAL SECTION.

Dimensions of Boiler No. 11.

Diameter of shell,	48	in.
Length between heads and length of tubes,	15	ft.
Number of tubes 3 inches outside diameter,	49	
Area of heating surface,	639	sq. ft.
Area of grate surface,	18	sq. ft.
Area through tubes,	2	sq. ft.
Ratio of heating surface to grate surface,	35.5	to 1
Ratio of grate to tube area,	8.9	to 1

Results of Tests. Boiler No. 11.

	Test No. 23.	Test No. 24.
Kind of fuel,	Anthracite broken.	Petroleum oil.
Manner of start and stop and kind of run,	Ordinary.	Ordinary.
Duration, ... hrs.	10	10
Fuel consumed, ... lbs.	986	1,365
Percentage of ash in coal, . . per cent.	15	—
Water evaporated, ... lbs.	8,493	15,787
Fuel per hour, ... lbs.	98.6	136.5
Fuel per hour per square foot of grate, . lbs.	5.7	—
Water per hour, ... lbs.	849.3	1,578.7
Water per hour per square foot of heating surface, ... lbs.	1.3	2.5
Horse-power developed, . . . H. P.	25.5	47.4
Boiler pressure, ... lbs.	59	62
Temperature of feed-water, . . . deg.	208	208
Water per pound of fuel, . . . lbs.	8.61	11.56
Water per pound of fuel from and at 212 degrees, ... lbs.	8.91	11.96
Water per pound of combustible from and at 212 degrees, ... lbs.	10.48	—

The tests on Boiler No. 11 had for an object the determination of the economy of crude petroleum, compared with anthracite coal. On the test with petroleum the damper was kept wide open, and the capacity of the boiler was regulated by varying the quantity of oil consumed. On the test with coal the damper was partially closed. During the oil test the quantity of steam used by the burner amounted to 15 per cent. of the total amount generated by the boiler, and the quantities given in the Table are corrected for the steam thus used.

The results of the tests show a higher evaporation per pound of fuel in the case of oil than in the case of coal, the increase being 34.2 per cent. The quantity of fuel required to evaporate a given quantity of water — say 30,000 pounds — from and at 212 degrees in the two cases, according to these results, is 3,367 pounds or 1.5 tons of coal, and 2,509 pounds or 386 gallons of oil. The coal at $5 per ton costs $7.50. The price of oil required to bring the cost of 386 gallons up to $7.50 is 1.9 cents per gallon. The employment of a liquid fuel serves to reduce to a minimum the labor attending the firing and care of a plant of boilers. A plant of 1,000 horse-power, in a mill running 10 hours per day, requires two firemen for day run, two helpers for wheeling coal, and one night hand for banking, removing ashes, and preparing the morning fires. The use of oil suitably arranged requires only one man, and dispenses with the labor of the remaining four. The saving in cost of labor thus realized represents about 10 per cent. of the cost of the fuel, assuming that coal is $5 per ton and 12 tons of coal are burned per day. This element in the problem should not be disregarded when making a comparison of the relative economy of the two kinds of fuel. In the case of these tests it makes the price of oil for the plant noted 2.18 cents per gallon to equal the performance of coal at $5 per ton.

Boiler No. 12.

Kind of boiler,	Horizontal return tubular.
Number used,	One.
Horse-power (basis 12 square feet),	Eighty-seven.
Age,	Eight months.

Boiler No. 12 is a horizontal tubular boiler, differing from that of the ordinary type in the arrangement of tubes. The general features of the boiler and the manner in which it is set are shown in the following cuts. The tubes are divided into two sections; one section serving to carry the products of combustion forward, as in the ordinary boiler, and the other serving to carry them backward to the chimney. The latter are placed centrally with reference to the former, and extend backward from the main tube sheet a distance of three feet, the extended portion being enclosed in a supplementary shell 34 inches in diameter, which forms an extension to the rear end of the boiler.

This arrangement of tubes reduces the area for draught in a given size of shell below what it would be in the boiler of the ordinary type, inasmuch as only about one-half of the tubes are effective in either direction. The ratio of the grate surface to the smallest tube area in the boiler under notice is 11.6 to 1.

The setting of the boiler is so arranged as to introduce air to the products of combustion as they leave the furnace. The air is supplied through passages which extend back and forth through the side walls, being finally discharged through perforated plates located in the sides of the furnace and in the top of the bridge wall.

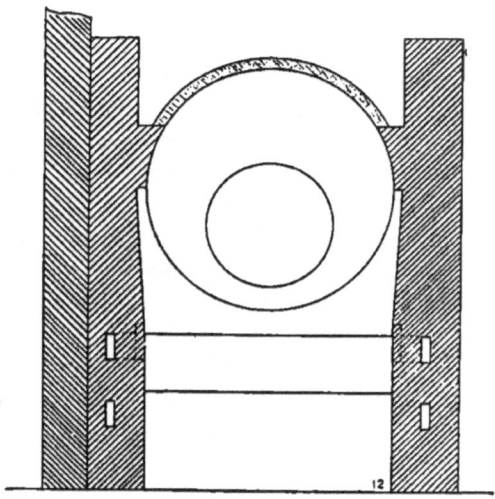

BOILER No. 12, CROSS SECTION THROUGH FURNACE.

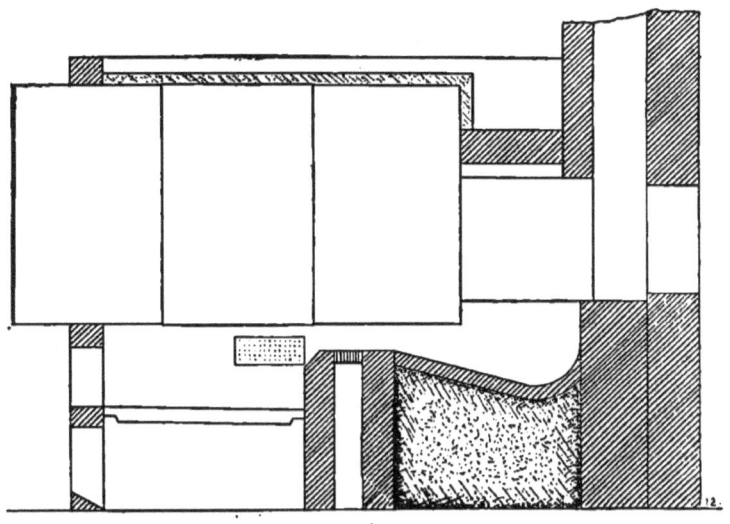

BOILER NO. 12, LONGITUDINAL SECTION.

Dimensions of Boiler No. 12.

Diameter of shell,	66 in.
Length of main shell between heads,	10 ft.
Number of tubes three inches outside diameter,	107
Length of 52 tubes,	10 ft.
Length of 55 tubes,	13 ft.
Area of heating surface,	1,041 sq. ft.
Area of grate surface,	24.7 sq. ft.
Least area for draught through tubes,	2.1 sq. ft.
Area through chimney,	2.3 sq. ft.
Height of chimney,	95 ft.
Width of air spaces and metal bars in grates,	3–8 in.
Distance of grate to shell,	23 in.
Distance of bridge wall to shell,	7 in.
Ratio of heating surface to grate surface,	42 to 1
Ratio of grate to tube area,	11.6 to 1

Results of Tests. Boiler No. 12.

	Test No. 25.	Test No. 26.	Test No. 27.	Test No. 28.
Kind of coal,	Lackawanna broken.	Lehigh broken.	George's Creek Cumberland.	2 parts Screenings, 1 part Geo.'s Creek Cumberland.
Manner of start and stop and kind of run,	Ordinary.	Ordinary.	Ordinary.	Ordinary.
Duration, . . . hrs.	10.7	10.2	10.7	11
Coal consumed, dry (including wood equivalent), lbs.	3,037	3,087	3,730	3,315
Percentage of ash, per cent.	12	10.1	6.6	16.5
Water evaporated, . lbs.	26,562	27,441	34,743	27,587
Coal per hour, . . lbs.	282.5	301.2	347	301.4
Coal per hour per square foot of grate, . . lbs.	11	12.2	14	12.2
Water per hour. . lbs.	2,470.9	2,677.2	3,231.9	2,507.9
Water per hour per square foot of heating surface, lbs.	2.4	2.6	3.1	2.4
Horse-power developed, H.P.	79.5	88.5	105.4	82
Boiler pressure, . lbs.	70	76.6	77.6	76.1
Temperature of feed-water, deg.	128	118	111	108
Temperature of escaping gases, . . . deg.	306	346	381	343
Draught suction, . . in.	0.28	0.28	0.28	0.28
Water per pound of coal, lbs.	8.75	8.89	9.31	8.32
Water per pound of coal from and at 212 degrees, lbs.	9.80	10.07	10.61	9.51
Water per pound of combustible from and at 212 degrees, . . . lbs.	11.13	11.20	11.37	11.40

NOTE. — The mixed fuel when fired contained 5 per cent. of moisture.

The tests on Boiler No. 12 had for an object the determination of the general economy of this particular form of boiler, and the relative economy produced by different kinds of fuel. The tests were made with a wide open damper and a constant maximum draught. They were thus capacity tests as well as. economy tests. The results obtained on Tests No. 25 and No. 26, made with anthracite coal, compare favorably with the best performance of horizontal tubular boilers. It is probable that even better results would have been obtained, if, on these tests, there had been no air admitted above the fuel. In point of economy the particular arrangement of tubes adopted appears to have been advantageous. The low temperature of the waste gases shows that the heating surface by this means was arranged so as to absorb the heat in an efficient manner. The favorable performance of the anthracite coals is borne out by the result of Test No. 28, made with the mixture. It is not so well borne out by the test of the Cumberland coal. Comparing the different results with that obtained on Test No. 25, made with Lackawanna coal, there is a gain for Lehigh coal of 2.8 per cent., a gain for Cumberland coal of 8.2 per cent., and a loss for the mixed fuel of 3 per cent., the comparisons being based on the coal results. These differences seem to be mainly due to the different percentages of ash, for the results based on combustible have an extreme variation of only 2.5 per cent. In the matter of capacity the Lackawanna coal gave a trifle less than the rated power, the Lehigh gained 11 per cent. over the Lackawanna, the Cumberland gained 33 per cent. over the same and the mixture 3 per cent.

The temperature of the escaping gases in the front connection, before passing into the central tubes was 325 degrees higher than that in the flue at the entrance to the chimney.

Boiler No. 13.

Kind of boiler,	Horizontal return tubular.
Number used,	One.
Horse-power (basis 12 sq. ft.),	Fifty-five.
Kind of coal,	Anthracite, Lehigh, Chestnut.
Age,	Sixteen years.

Boiler No. 13 is the same boiler as the return tubular boiler given under the head of Boiler No. 1, with the exception that the superheater had been removed. The general form of the boiler and the arrangement of the setting are shown in longitudinal section in the following cut. The tests here given (Nos. 29 and 30), were made one year later than Tests Nos. 1 and 2.

BOILER No. 13.

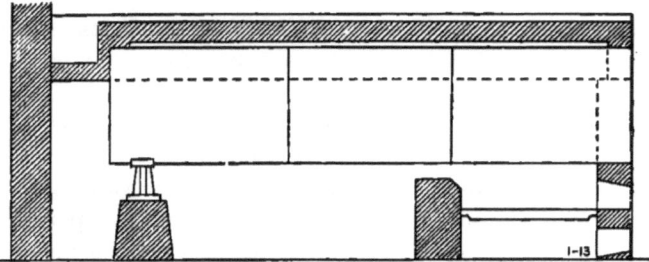

BOILER No. 13, LONGITUDINAL SECTION.

Dimensions of Boiler No. 13.

Diameter of shell,	48 in.
Length between heads and length of tubes,	16 ft.
Number of tubes three inches outside diameter,	48
Area of heating surface,	669 sq. ft.
Area of grate surface,	19.9 sq. ft.
Area through tubes,	2 sq. ft.
Area through flue,	1.9 sq. f
Height of chimney,	77 ft.
Width of air spaces and metal bars in grates,	Air 3-8 in., metal 5-8 1
Distance of grate to shell,	19 in.
Distance of flat bridge wall to shell,	6 in.
Ratio of heating surface to grate surface,	33.7 to 1
Ratio of grate to tube area,	10 to 1

Results of Tests, Boiler No. 13.

	Test No. 29.	Test No. 30.
Conditions,	Damper wide open.	Damper partly open.
Manner of start and stop and kind of run,	Ordinary.	Ordinary.
Duration, hrs.	11	11
Coal consumed, (including wood equivalent,) lbs.	2,647	1,430
Percentage of ash, per cent.	16 4	17.5
Water evaporated, lbs.	19,653	10,731
Coal per hour, lbs.	240 6	130
Coal per hour per square foot of grate, lbs.	12.1	6.5
Water per hour, lbs.	1,786.6	975.6
Water per hour per square foot heating surface, lbs.	2.7	1.5
Horse-power developed, H. P.	62.1	33.9
Boiler pressure, lbs.	71.9	69.5
Temperature of feed-water, deg.	40	40
Temperature of escaping gases, deg.	444	356
Water per pound of coal, lbs.	7.42	7.50
Water per pound of coal from and at 212 degrees, lbs.	8.99	9.08
Water per pound of combustible from and at 212 degrees, lbs.	10.76	11 01

The object of the tests on Boiler No. 13 was to determine the effect of two widely different rates of combustion. In Test No. 29, the rate was 12.1 pounds per square foot of grate per hour, and in Test No. 30 it was 6.5 pounds, one giving a slight excess above the nominal power, and the other a much lower capacity. The difference in the economic results of the two tests in favor of the slow rate of combustion is one per cent. based on coal, and 2.3 per cent. based on combustible. There is a noticeable difference in the two tests in the temperature of the waste gases. The excessive temperature with the high rate of combustion, taken in connection with the age of the boiler, gives evidence of the presence of scale on the heating surfaces. Had the surface been of sufficient extent, or in such a condition as to absorb this heat, the comparison between the two tests would undoubtedly have been in favor of rapid combustion.

Boiler No. 14.

Kind of boiler,	Horizontal direct tubular.
Number used,	One.
Horse-power (basis 12 square feet),	Fifty-two.
Kind of coal,	Anthracite, Lehigh, Chestnut.
Age,	Sixteen years.

Boiler No. 14 is the same boiler as that referred to as the direct tubular boiler under the head of Boiler No. 1, the tests being made one year later. The general arrangement of the boiler and the manner in which it is set are shown in longitudinal section in the following cut.

In this boiler the products of combustion pass upward from the furnace into a combustion chamber situated in the front end of the shell, then forward through the tubes, then backward under the shell to the furnace wall, and finally upward through two side flues and over the top of the rear part of the shell to the chimney flue. The upper part of the shell last traversed furnishes steam-heating surface. This surface is probably inefficient, owing to the deposit of soot which naturally covers it. The boiler is fitted with a steam dome.

BOILER No. 14.

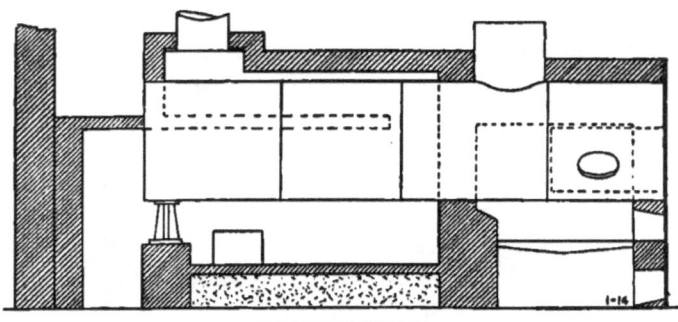

BOILER NO. 14, LONGITUDINAL SECTION.

Dimensions of Boiler No. 14.

Diameter of shell,	48 in.
Length between heads and length of tubes,	14 ft.
Number of tubes 3 inches outside diameter,	45
Length of combustion chamber,	38 in.
Inside diameter of combustion chamber,	42 in.
Diameter of two openings to combustion chamber,	14 in.
Area of water-heating surface,	586 sq. ft.
Area of steam-heating surface,	43 sq. ft.
Area of grate surface,	21.2 sq. ft.
Area through tubes,	1.9 sq. ft.
Area through flue,	2.2 sq. ft.
Width of air spaces and metal bars in grates,	Air 3-8 in., metal 5-8 in.
Distance of grate to shell,	17 in.
Height of chimney,	77 ft.
Ratio of water-heating surface to grate surface,	27.7 to 1
Ratio of grate to tube area,	11.4 to 1

Results of Tests, Boiler No. 14.

	Test No. 31.	Test No. 32.
Conditions,	Damper wide open.	Damper partly closed.
Manner of start and stop and kind of run,	Ordinary.	Ordinary.
Duration, ... hrs.	11.2	10.7
Coal consumed, (including wood eqivalent,) . . . . lbs.	2,731	1,367
Percentage of ash, . . . per cent.	15.7	15.5
Water evaporated, ... lbs.	19,891	10,388
Coal per hour . . . . . lbs.	242.8	127.2
Coal per hour per square foot of grate, lbs.	11.4	6
Water per hour, ... lbs.	1,768.1	966.3
Water per hour per square foot of water-heating surface, ... lbs.	3	1.6
Horse power developed, . . H. P.	61.4	33.5
Boiler pressure, ... lbs.	67.7	65.2
Temperature of feed-water, . deg.	40	40
Temperature of escaping gases, . deg.	476	380
Water per pound of coal, . . lbs.	7.28	7.60
Water per pound of coal from and at 212 degrees, ... lbs.	8.82	9.19
Water per pound of combustible from and at 212 degrees, ... lbs.	10.46	10.88

The tests on Boiler No. 14 like those on No 13, were made to determine the effect produced by two widely different rates of combustion. In Test No. 31 the rate was 11.4 pounds of coal per square foot of grate per hour, and in Test No. 32 it was 6 pounds, one giving a somewhat larger amount of power than the rated boiler power, and the other a much lower amount. The economical result obtained with the low capacity is the better of the two to the extent of 4.2 per cent. There is a noticeable difference in the two tests in the temperature of the escaping gases, one being 476 and the other 380 degrees. To this difference is evidently due the comparatively low economy produced by the rapid combustion. Comparing these results with those obtained from Boiler No. 13, they are inferior, as might be expected from the difference in the quantity of heating surface, and the resulting higher temperature of the waste gases. There is 20 per cent. less heating surface in proportion to grate in this boiler, than in Boiler No. 13.

Boiler No. 15.

Kind of boiler,	Horizontal return tubular.
Number used,	One.
Horse-power (basis 12 square feet),	Seventy-four.
Kind of coal,	Anthracite, Lehigh, Chestnut.
Age,	One year.

Boiler No. 15 is similar in general features to Boiler No. 9, being provided with a water leg front, which takes the place of the ordinary cast iron front with brick lining. The general arrangement of the boiler and the manner in which it is set are shown in the following cuts. It is to be noted that the space behind the bridge wall is unusually deep, extending a distance of 7 feet below the shell, and one of the side walls is exposed on the outside a corresponding distance.

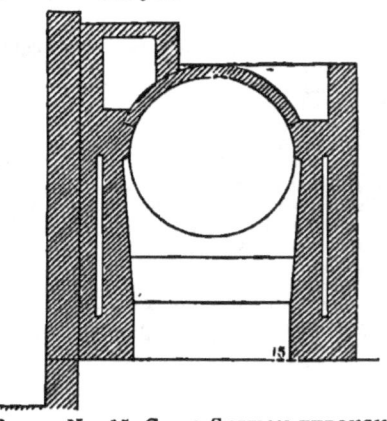

BOILER No. 15, CROSS SECTION THROUGH FURNACE.

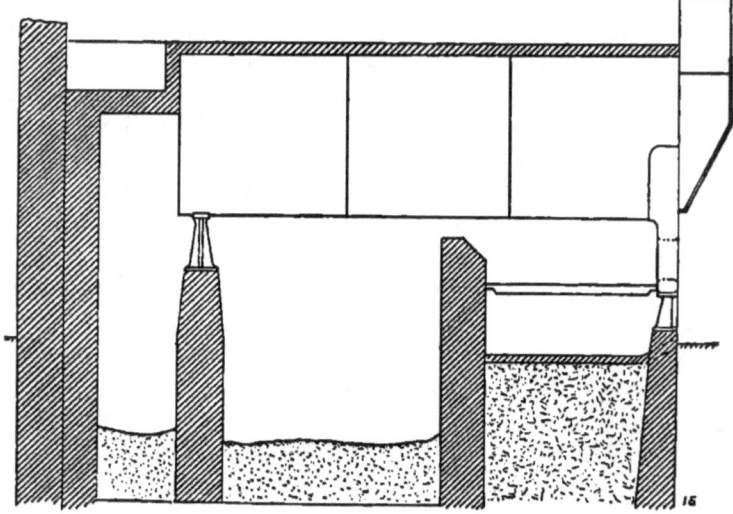

BOILER NO. 15, LONGITUDINAL SECTION.

Dimensions of Boiler No. 15.

Diameter of shell,	60 in.
Length between heads and length of tubes,	15 ft.
Number of tubes 3 inches outside diameter,	66
Area of heating surface,	890 sq. ft.
Area of grate surface,	24.1 sq. ft.
Area through tubes,	2.7 sq. ft.
Area through flue,	2.4 sq. ft.
Height of chimney,	77 ft.
Width of air spaces and metal bars in grates, Air 3-8 in., metal 5-8 in.	
Distance of grate to shell,	23 in.
Distance of flat bridge wall to shell,	7 in.
Ratio of heating surface to grate surface,	37 to 1
Ratio of grate to tube area,	10 to 1

Results of Tests, Boiler No. 15.

	Test No. 33.	Test No. 34.
Conditions,	Damper wide open.	Damper partly closed.
Manner of start and stop and kind of run,	Ordinary.	Ordinary.
Duration, hrs.	12	11
Coal consumed (including wood equivalent), lbs.	4,099	2,659
Percentage of ash, per cent.	14.4	14.2
Water evaporated, lbs	30,326	20,244
Coal per hour, lbs	341.6	241.7
Coal per hour per square foot of grate, lbs	14.2	10
Water per hour, lbs.	2,527.2	1,840.4
Water per hour per square foot of heating surface, lbs.	2.8	2.1
Horse-power developed, H. P.	87.8	64
Boiler pressure, lbs.	73.8	70
Temperature of feed-water, deg.	40	40
Temperature of escaping gases, deg.	365	350
Water per pound of coal, lbs.	7.40	7.61
Water per pound of coal from and at 212 degrees, lbs.	8.97	9.22
Water per pound of combustible from and at 212 degrees, lbs.	10.47	10.75

The tests on Boiler No. 15, together with those made on Boilers No. 13 and 14, form a group of tests, the object of which was to determine the relative economy of working to maximum capacity and to medium capacity, in the case of three boilers of somewhat different types. All used coal from the same cargo. In the case of Boiler No. 15, as in those of No.

13 and No. 14 already noted, there is an appreciable difference of economy in favor of the medium capacity. Here it is about three per cent. There is a notably low temperature of the escaping gases compared with that found in the other two boilers worked at maximum capacity, due, no doubt, in part, to the relatively large extent of heating surface, and in part to the cleanliness of the surfaces which presumably existed in the new boiler. Still, this favorable indication is accompanied by a comparatively low evaporative result. The large space behind the bridge and the greater radiating surfaces thus exposed does not appear to be adequate to produce an unfavorable effect of much consequence, but there appears no other cause. It is noted that a higher rate of combustion is secured in Boiler No. 15 than in the others. This may be attributed to the newer and tighter condition of the flues leading to the chimney. One chimney of ample area served for all three boilers.

A series of tests was made on Boiler No. 15 to compare the economy of different fuels. These are numbered from 35 to 39 inclusive. With the exception of Test No. 39 made with mixed fuel, none of these are capacity tests. On all the tests except this one, the damper was partially closed. The results are given in the appended table.

Taking the results with chestnut coal as a basis of comparison, and figuring on the evaporation per pound of coal, the pea coal lost 9 per cent.; the broken coal gained 4.2 per cent.; the Cumberland coal gained 19.3 per cent., and the mixed fuel gained 6.2 per cent.

Considering the cost of these fuels given below, as quoted at the time of the tests (per ton of 2,240 pounds),

Chestnut,	$5 50
Pea,	4 60
Broken,	5 85
Pea and Dust,	2 75
Cumberland,	6 50

the total value of the fuel used in evaporating a given quantity of water, say 30,000 pounds from and at 212 degrees, according to the results here obtained, is as follows:

Results of Tests, Boiler No. 15 (continued).

		Test No. 35.	Test No. 36.	Test No. 37.	Test No. 38.	Test No. 39.
Kind of fuel,		Honeybrook Lehigh Chestnut. Ordinary.	Honeybrook Lehigh Pea. Ordinary.	Honeybrook Lehigh Broken. Ordinary.	George's Creek Cumberland.* Ordinary.	Pea and Dust 2 pts., Cumberland 1 pt.* Ordinary.
Manner of start and stop and kind of run,		11	11	10.75	10.75	10.75
Duration,	hrs.	2,659	2,869	3,419	2,369	2,792
Coal consumed, dry (including wood equivalent),	lbs.	14.2	15.8	10.5	6.6	11.4
Percentage of ash,	per cent.	20,244	19,919	27,089	21,520	22,537
Water evaporated,	lbs.	241.7	260.9	318	220.4	259.7
Coal per hour,	lbs.	10	10.8	13.2	9.1	10.7
Coal per hour per square foot of grate,	lbs.	1,840.4	1,810.8	2,519.9	2,001.9	2,096.5
Water per hour,	lbs.	2.1	2	2.8	2.2	2.4
Water per hour per square foot of heating surface,	H.P.	64	62.8	87.7	69.6	72.9
Horse-power developed,	lbs.	70	70	72.6	69.9	71.6
Boiler pressure,	deg.	40	40	39	39	39
Temperature of feed-water,	deg.	350	348	372	387	387
Temperature of escaping gases,	lbs.	7.61	6.94	7.92	9.08	8.07
Water per pound of coal,	lbs.	9.22	8.40	9.61	11	9.79
Water per pound of coal from and at 212 degrees,						
Water per pound of combustible from and at 212 degrees,	lbs.	10.75	9.98	10.74	11.78	11.06

* These coals were moist when fired.

Chestnut, ...	$8 00
Pea, ...	7 32
Broken, ...	8 15
Cumberland, ...	7 91
Mixed fuel, ...	5 48

According to this basis of comparison, which may be called the commercial basis, the cost with pea coal is reduced $0.32 or 8.5 per cent.; that with broken coal is increased $0.15, or 1.9 per cent.; that with Cumberland is reduced $0.09, or 1.1 per cent.; and that with the mixture is reduced $2.52, or 31.5 per cent., these being all compared with the cost when using chestnut coal. Although this comparison does not apply to the present time, when prices of coal are altogether different, the figures are suggestive as to the influence which cost has upon the actual value of a given kind of coal.

The varying amounts of ash which different sizes of anthracite coal, and the different classes of other coals contain, is exhibited by these tests. The pea coal contained the largest quantity, viz: 15.8 per cent.; and the broken coal the smallest for the anthracite class, viz: 10.5 per cent; while the bituminous George's Creek Cumberland gave 6.6 per cent; and the mixed fuel 11.4 per cent. The last is low for this class of fuel and an indication of good quality. There is a noticeable difference in the effect of the various fuels on the temperature of the escaping gases. The Cumberland and mixed fuels gave a higher temperature than that produced by the anthracite coals. This condition is almost always observed in fuels made up in whole or in part of bituminous coal.

Another series of tests was made on Boiler No. 15, the object of which was to determine the effect upon the economy of different fuels produced by admitting air at the bridge wall. These are numbered from 40 to 45 inclusive, and the data and results are given in the appended table. The method employed in supplying air, consisted in introducing it through a pipe 7 inches in diameter running through the side wall, at a distance of 26 inches below the top of the bridge wall. The pipe entered a chamber formed by building a new wall a few inches behind the bridge and covering the top of the intervening

BOILER TESTS.

Results of Tests, Boiler No. 15 (concluded).

	Test No. 40.	Test No. 41.	Test No. 42.	Test No. 43.	Test No. 44.	Test No. 45.
Kind of coal,	Cumberland, George's Creek.*	Cumberland, George's Creek.*	Pea and Dust 2 parts, Cumberland 1 part.*	Pea and Dust 2 parts, Cumberland 1 part.*	Anthracite Lehigh Broken.	Anthracite Lehigh Broken.
Conditions as to admission of air at bridge wall,	Air admitted.	No air admitted.	Air admitted.	No air admitted.	Air admitted.	No air admitted.
Manner of start and stop and kind of run.	Ordinary.	Ordinary.	Ordinary.	Ordinary.	Ordinary.	Ordinary.
Duration, ... hrs.	10.75	10.75	10.75	11	11	10.75
Coal consumed, dry (including wood equivalent), lbs.	2,369	2,556	2,792	2,874	3,236	3,419
Percentage of ash, per cent.	6.6	6.4	11.4	13.8	10	10.5
Water evaporated, lbs.	21,520	21,892	22,537	23,685	25,597	27,089
Coal per hour, . lbs.	220.4	237.3	259.7	261.3	294.2	318
Coal per hour per square foot of grate, lbs.	9.1	9.4	10.7	10.8	12.2	13.2
Water per hour, lbs.	2,001.9	2,036.5	2,096.5	2,153.2	2,327	2,519.9
Water per hour per square foot of heating surface, lbs.	2.2	2.3	2.4	2.4	2.6	2.8
Horse-power developed, H. P.	69.6	70.8	72.9	74.8	81.9	87.7
Boiler pressure, lbs.	69.9	68.1	71.6	72	68.2	72.6
Temperature of feed-water, deg.	39	39	39	39	39	39
Temperature of escaping gases, deg.	387	372	387	362	389	372
Water per pound of coal, lbs.	9.08	8.57	8.07	8.24	7.91	7.92
Water per pound of coal from and at 212 degrees, lbs.	11	10.39	9.79	9.99	9.59	9.61
Water per pound of combustible from and at 212 degrees, lbs.	11.78	11.09	11.06	11.60	10.65	10.74

* These coals were moist when fired.

space with a perforated iron plate, which was placed on a level with the top of the bridge. A damper located in the pipe, was wide open when Cumberland coal was used and about half-way open with the other fuels.

The effect upon the economic result produced by the admission of air at the bridge was quite marked in the cases of the Cumberland coal and mixed fuel, but not so in that of the anthracite coal. The evaporation per pound of combustible was increased 6.2 per cent. with Cumberland coal, but it was decreased 4.7 per cent. with the mixed fuel. There was little difference produced with anthracite coal.

In every case the admission of air was accompanied by an increased temperature of the escaping gases and a decreased amount of power developed. These differences are small, but they are always in the same direction.

The character of the combustion, as seen by the eye of an observer looking through a peek hole into the space behind the bridge wall, was always improved by the admission of air.

The effect of admitting air upon the Cumberland coal was to reduce the quantity and density of the smoke discharged at the top of the chimney. There was entire absence of smoke $\frac{30}{100}$ of the time when air was admitted, and $\frac{20}{100}$ of the time when air was not admitted. There was little smoke in either case with mixed fuel.

There was no appreciable difference in the quantity of soot deposited in the tubes, as shown by examination at the front ends after each test, whether air was admitted at the bridge or not.

The draught suction at the bottom of the chimney was ordinarily $\frac{7}{16}$ of an inch, expressed in terms of water pressure. In the boiler flue it was $\frac{3}{8}$ of an inch with wide open damper, and in the space behind the bridge wall $\frac{3}{16}$ of an inch.

Boiler No. 16.

Kind of boiler,	Horizontal return tubular.
Number used,	One.
Horse-power (basis 12 square feet),	Seventy-five.
Age,	Two years.

Boiler No. 16 is of the ordinary horizontal return tubular type, with a flush front, the setting of which is shown in longitudinal section in the cut. This boiler was one of a large battery, and during the progress of the tests the boiler on each side was in daily use.

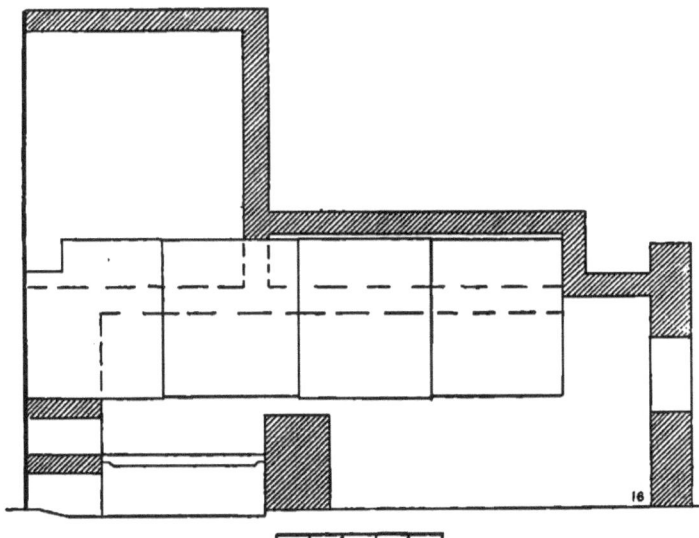

BOILER No. 16, LONGITUDINAL SECTION.

Dimensions of Boiler No. 16.

Diameter of shell,	60 in.
Length between heads and length of tubes,	15 ft.
Number of tubes three inches outside diameter,	70
Area of heating surface,	900 sq. ft.
Area of grate surface,	26.7 sq. ft.
Area through tubes,	2.9 sq. ft.
Area through flue,	3 sq. ft.
Height of chimney,	120 ft.
Width of air spaces and metal bars in grates,	3-8 in.
Distance of grate to shell,	20 in.
Distance of flat bridge to shell,	6 in.
Ratio of heating surface to grate surface,	33.8 to 1
Ratio of grate surface to tube area,	92. to 1

BOILER No. 16.

Results of Tests, Boiler No. 16. (Average of two.)

	Test No. 46.	Test No. 47.	Test No. 48.
Kind of coal,	Anthracite Stove.	Anthracite Chestnut No. 2.	2 parts Pea and Dust, 1 part Clearfield.
Manner of start and stop and kind of run,	Ordinary.	Ordinary.	Ordinary.
Duration, ... hrs.	10.7	11	10.7
Coal consumed, dry (including wood equivalent), . lbs.	2,661	2,700	2,789
Percentage of ash, . per cent.	14.7	12.8	14
Water evaporated, . . lbs.	21,213	21,350	22,030
Coal per hour, . . . lbs.	247.7	245.5	259.4
Coal per hour per square foot of grate, ... lbs.	9.3	9.6	9.7
Water per hour, . . lbs.	1,974.4	1,941	2,049.3
Water per hour per square foot of heating surface, . lbs.	2.2	2.2	2.3
Horse-power developed, H. P.	66.9	65.8	69.3
Boiler pressure, . . lbs.	65.8	64.8	67.9
Temperature of feed-water, deg.	67	66.5	71
Temperature of escaping gases, deg.	314	312	326
Number of firings, . . .	20	37	39
Number of times slice bar was used, ...	1	1	12
Position of damper, . . .	¼ open.	¼ open.	½ open.
Water per pound of coal, . lbs.	7.97	7.90	7.90
Water per pound of coal from and at 212 degrees, . lbs.	9.42	9.34	9.31
Water per pound of combustible from and at 212 degrees, lbs.	11.06	10.72	10.85

NOTE.— The mixed fuel when fired contained 4 1-2 per cent. of moisture.

The tests on Boiler No. 16 had for an object the determination of the relative economy of three different kinds of coal, viz: anthracite stove coal, anthracite chestnut No. 2, and a mixture of two parts pea and dust and one part Clearfield bituminous. On the first two tests the damper was one-fourth open; on the test with the mixed fuel the damper was open half way, and in all cases the damper was kept in a fixed position. The evaporative results of the three tests, based on coal, are almost identical. The various results based on the cost of fuels required to evaporate 30,000 pounds of water from and at 212 degrees, according to the prices which ruled at the time the tests were made, are as follows:

	Stove.	Chestnut No. 2.	Mixture.
Cost per ton of 2,240 pounds, . . .	$5 40	$4 50	$3 70
Cost of fuel for 30,000 pounds of steam, .	7 63	6 42	5 33

A comparison of these figures shows that the cost of making a given amount of steam with chestnut No. 2 coal was 16 per cent. less than with stove coal, and the cost with the mixture was 30 per cent. less.

It is noticeable in the case of the test with mixed fuel that the labor of firing was greater than that with either of the other coals. There were 39 firings with the mixed fuel, against 20 with the stove coal and 37 with chestnut No. 2 coal, while the number of times the slice bar was used was 12 with mixed fuel, and only one in each case with the others.

The low temperature of the escaping gases is noticeable in these tests. This is an indication of economical work which the favorable character of the evaporative results plainly bears out.

Boiler No. 17.

Kind of boiler, ... Horizontal return tubular.
Number used, ... Three.
Horse-power (collective basis 12 square feet), One hundred and twenty-nine.
Age, ... Several years.

Boiler No. 17 embraces three ordinary horizontal tubular boilers set in brick work in the manner shown in the following cuts. Two of these boilers are set with a furnace common to both, and in all three the escaping gases, after leaving the smoke arches in front, pass over the tops of the shells. The three boilers are end boilers of the battery, and the next boiler in the set was in daily operation during the progress of the tests.

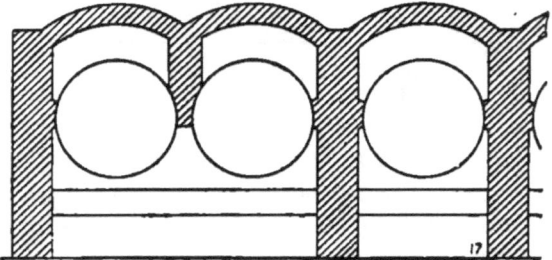

BOILER NO. 17, CROSS SECTION THROUGH FURNACE.

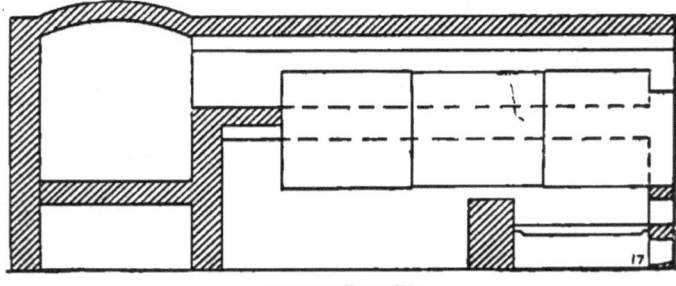

BOILER NO. 17, LONGITUDINAL SECTION.

Dimensions of Boiler No. 17.

Diameter of shell,	48	in.
Length between heads and length of tubes,	12	ft.
Number of tubes (collective) 3 inches outside diameter,	147	
Area of water-heating surface,	1,548	sq. ft.
Area of steam-heating surface,	185	sq. ft.
Area of grate surface,	58.5	sq. ft.
Area through tubes,	6.1	sq. ft.
Height of chimney,	140	ft.
Width of air spaces and metal bars in grates,	Air 7-16 in., metal 3-8 in.	
Distance of grate to shell,	16	in.
Distance of flat bridge to shell,	5	in.
Ratio of water-heating surface to grate surface,	26.5	to 1
Ratio of grate to tube area,	9.6	to 1

Results of Tests, Boiler No. 17.

	Test No. 49.	Test No. 50.	Test No. 51.
Kind of coal,	Anthracite White Ash Broken.	Anthracite Pea.	44 parts Pea and Dust, 37 parts Nova Scotia Culm.
Manner of start and stop and kind of run,	Ordinary.	Ordinary.	Ordinary.
Duration, ... hrs.	10.2	10.5	10.5
Coal consumed, dry (including wood equivalent), . lbs.	7,848	6,584	7,595
Percentage of ash, . per cent.	10.1	20.9	17.9
Water evaporated, . . lbs.	56,715	45,071	47,452
Coal per hour, . . . lbs.	756.6	627	723.3
Coal per hour per square foot of grate, ... lbs.	12.9	11.7	13.6
Water per hour, . . . lbs.	5,532.6	4,292.4	4,519.2
Water per hour per square foot of water-heating surface, lbs.	3.6	2.8	2.9
Horse-power developed, H. P.	192.3	149.2	157.1
Boiler pressure, . . lbs.	73.7	74.2	73
Temperature of feed-water, deg.	39	39	39
Temperature of escaping gases, deg.	455	448	460
Draught suction, . . . in.	0.11	0.12	0.28
Number of firings, . . .	23	27	34
Number of times using slice bar or hoe, ...	6	5	24
Water per pound of coal, . lbs.	7.23	6.84	6.24
Water per pound of coal from and at 212 degrees, . lbs.	8.77	8.29	7.57
Water per pound of combustible from and at 212 degrees, lbs.	9.75	10.63	9.34

NOTE.—The pea coal when fired contained 5 per cent. of moisture and the mixture 10 per cent.

The tests on Boiler No. 17 had for a main object the determination of the relative economy of three different kinds of fuel, viz: anthracite broken coal, anthracite pea coal, and a mixture of pea and dust and Nova Scotia culm. In general, the results do not show a high degree of economy. In view of the small ratio of heating surface to grate surface, the relatively large amount of power developed, compared with the nominal power, and the resulting high temperature of the waste gases, a somewhat unfavorable result would be expected. Comparing the performance of the different fuels, the evaporation with pea coal is 5.5 per cent. less than that obtained with broken coal, and the result obtained with the mixed fuel is 13.7 per cent. less than the same figure, these being based on the evaporation per pound of coal. Basing the results on the cost of fuel required to produce 30,000 pounds of steam from and at 212 degrees, according to the prices which ruled at the time of the tests, the figures are as follows:

	Broken.	Pea.	Mixture.
Cost per ton of 2,240 pounds,	$6 10	$4 75	$3 85
Cost for 30,000 pounds of steam,	9 32	7 64	6 81

In this comparison the economy of pea coal is 18 per cent. over broken coal, and that of the mixture 26.9 per cent. over the same fuel.

It is to be noted that the amount of draught required on the test with mixed fuel was 0.28 inches measured in water pressure, while that required on the other tests was 0.11 and 0.12 inches respectively. The labor of firing with mixed fuel was much greater than in the case of the others, both the number of firings and the number of times that the slice bar was used being relatively large.

Boiler No. 18.

Kind of boiler,	Horizontal return tubular.
Number used,	One.
Horse-power (basis 12 sq. ft.).	Fifty-three.
Age,	Six months.

Boiler No. 18 is a horizontal tubular boiler with a special arrangement of setting, the general features of which are shown in longitudinal section in the following cut. A furnace is placed under each end of the boiler, located in the same manner with reference to the boiler as the furnace in the common horizontal boiler. A damper is provided in each front connection for the control of the flue gases, and another above each furnace at the entrance to the chamber in front of the tubes, to control the passage of the products of combustion into the boiler. The method of operation consists in firing the furnaces alternately. The products of combustion from either furnace are made to pass over the burning coal in the other furnace, whenever either one is supplied with fresh coal. Just previous to each firing the dampers are changed so that the desired result may be secured. The fire in the secondary furnace serves to heat and ignite the unburned gases formed in the first.

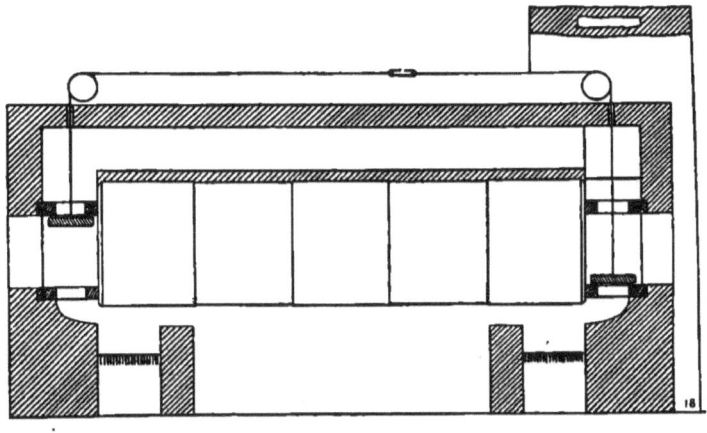

BOILER NO. 18, LONGITUDINAL SECTION.

Dimensions of Boiler No. 18.

Diameter of shell,	48	in.
Length between heads and length of tubes,	15	ft.
Number of tubes 3 inches outside diameter,	49	
Area of heating surface,	640	sq. ft.
Area of grate surface (total of two furnaces),	11.7	sq. ft.
Area through tubes,	2	sq. ft.
Height of chimney,	50	ft.
Width of air spaces and metal bars in grates,	5-8	in.
Distance of grate to shell,	18	in.
Distance of bridges to shell,	7	in.
Ratio of heating surface to grate surface,	54.4	to 1
Ratio of grate surface to tube area,	5.9	to 1

Results of Tests, Boiler No. 18.

	Test No. 52.	Test No. 53.
Kind of coal,	Bituminous Cumberland.	Anthracite Lehigh, broken.
Manner of start and stop and kind of run,	Ordinary.	Ordinary.
Duration, ... hrs.	10.5	11
Coal consumed, dry (including wood equivalent), ... lbs.	2,368	1,587
Percentage of ash, . . . per cent.	8.5	13.5
Water evaporated, ... lbs.	19,734	11,496
Coal per hour, ... lbs.	225.5	144.3
Coal per hour per square foot of grate, lbs.	19.3	12.5
Water per hour, ... lbs.	1,879.4	1,045.1
Water per hour per square foot of heating surface, ... lbs.	2.9	1.6
Horse-power developed, . . H. P.	65.3	36.3
Boiler pressure, ... lbs.	50.1	50.9
Temperature of feed-water, . . deg.	41	41
Temperature of escaping gases, . deg.	472	354
Draught suction, ... in.	0.25	0.12
Water per pound of coal, . . lbs.	8.33	7.24
Water per pound of coal from and at 212 degrees, ... lbs.	10.00	8.69
Water per pound of combustible from and at 212 degrees, . . . lbs.	10.93	10.05

The tests on Boiler No. 18 had for an object the determination of the economy produced by alternate firing in the double furnace system which was here employed. Two kinds of coal were used, viz: Cumberland coal on test No. 52, and anthracite Lehigh on Test No. 53. The Cumberland coal gave an evaporation of 10.93 pounds of water from and at 212 degrees per pound of combustible, and the boiler developed somewhat more than its rated capacity. This result is much below the highest obtained from boilers which are set and operated in the usual manner. The cause of the low performance may be attributed in part to the somewhat high degree of waste heat escaping to the chimney, the temperature of the gases being 472 degrees, though there does not appear to be a sufficient loss in this alone to wholly account for it. It is probable that the use of a furnace at each end of the boiler, and the arrangement of flues required in this system, is attended with a greater loss from radiation than ordinarily occurs, and that this loss more than offsets the advantage which might otherwise be obtained from the system. The quantity of smoke issuing from the chimney with Cumberland coal was materially less than occurs with the ordinary boiler setting, and this is an indication that the system exerts a favorable effect in securing more perfect combustion with bituminous coal. The test with anthracite coal also gave a low result, a fact which points even more plainly to the probability of excessive loss by radiation from the furnace walls, as noted. Here the economy which ordinarily attends a high rate of combustion and low flue temperature was not realized.

Boiler No. 19.

Kind of boiler,	Horizontal return tubular.
Number used,	Four.
Horse-power (collective, basis 12 sq. feet),	Two hundred and seventy-five.
Kind of coal,	Bituminous, Cumberland.

Boiler No. 19 embraces four horizontal return tubular boilers, set in one battery of brick work, the general features of which are shown in the cuts of Boiler No. 5. Compared with ordinary practice, the boiler has short tubes and large grates, and, as a consequence, the ratio of heating surface to grate surface is somewhat low, being 29.4 to 1.

Dimensions of Boiler No. 19.

Diameter of shell,	60	in.
Length between heads and length of tubes,	12	ft.
Number of tubes (collective) 3 inches outside diameter,	336	
Area of heating surface,	3,534	sq. ft.
Area of grate surface,	120	sq. ft.
Area through tubes,	13.8	sq. ft.
Area through flue,	15.9	sq. ft.
Height of chimney,	123	ft.
Width of air spaces and metal bars in grates,	Air 3-8 in., metal 7-16 in.	
Distance of grate to shell,	24	in.
Distance of flat bridge to shell,	9	in.
Ratio of heating surface to grate surface,	29.4 to 1	
Ratio of grate surface to tube area,	8.7 to 1	

Results of Tests, Boiler No. 19.

	Test No. 54.
Manner of start and stop,	Thin fire.
Kind of run,	Continuous.
Duration,	10.2 hrs.

BOILER No. 19.

Coal consumed, dry,	13,293	lbs.
Percentage of ash,	8.7	per cent.
Water evaporated,	113,115	lbs.
Coal per hour,	1,307.6	lbs.
Coal per hour per square foot of grate,	10.9	lbs.
Water per hour,	11,125.6	lbs.
Water per hour per square foot of heating surface,	3.1	lbs.
Horse-power developed,	380	H. P.
Boiler pressure,	91	lbs.
Temperature of feed-water,	117	deg.
Temperature of escaping gases,	530	deg.
Draught suction,	0.31	in.
Water per pound of coal,	8.50	lbs.
Water per pound of coal from and at 212 degrees,	9.67	lbs.
Water per pound of combustible from and at 212 degrees,	10.60	lbs.

NOTE.— The coal when fired contained 3 per cent. of moisture.

The test on Boiler No. 19 shows the performance of Cumberland coal in boilers which were not proportioned and operated in a manner to give the best results. The power developed is nearly 40 per cent. above the nominal capacity. The rate of combustion, though only 10.9 pounds per square foot of grate per hour, is high for a boiler having the relatively small amount of heating surface here shown. The temperature of the gases is excessive, being 530 degrees. The percentage of ash is high for Cumberland coal, and indicates some inferiority in the quality of the fuel. The evaporative result, as might be expected from these unfavorable conditions, exhibits a low degree of economy.

Boilers No. 20 and No. 21.

Kind of boilers,	Horizontal return tubular.
Number used, each,	One.
Horse-power each (basis 12 square feet),	Eighty.
Kind of coal, each,	{ Three parts pea and dust, one part Cumberland.
Age,	No. 20, two yrs., No. 21, one yr.

Boilers No. 20 and 21 embrace two horizontal tubular boilers set in one battery of brick work. They are provided with independent flues leading to the chimney, the arrangement of which is shown in the front elevation given in the following cut. The general arrangement of the brick setting of each

124   BOILER TESTS.

boiler is the same as that shown in the cuts of Boiler No. 6. The two boilers are identical, both as to size and general manner of setting, with the exception that No. 20 is provided with perforated plates at the sides of the furnace and top of the bridge wall, for admitting air above the fuel, the air first passing back and forth through ducts in the side walls. Boiler No. 21 is provided with the ducts in the same manner as No. 20, but the entrance to these is closed, and no outlets are provided opening into the furnace.

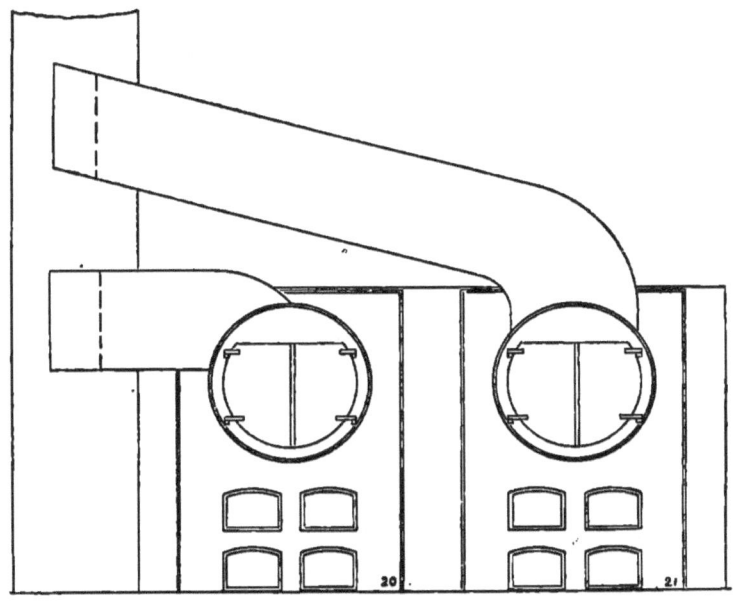

BOILERS NO. 20 AND NO. 21, FRONT ELEVATION.

Dimensions of Boilers No. 20 and No. 21.

Diameter of each shell,	60 in.
Length between heads and length of tubes,	15 ft.
Number of tubes 3¼ inches outside diameter,	64
Area of heating surface,	959 sq. ft.
Area of grate surface,	25.8 sq. ft.
Area through tubes,	3.7 sq. ft.

BOILERS No. 20 and No. 21.

Area through flue,	3 sq. ft.
Height of chimney,	75 ft.
Width of air spaces and metal bars in grates,	3-8 in.
Distance of grate to shell,	26 in.
Distance of flat bridge wall to shell,	9 in.
Ratio of heating surface to grate surface,	37.1 to 1
Ratio of grate surface to tube area,	7 to 1

Results of Tests, Boilers No. 20 and No. 21. (Average of 2 runs.)

	Boiler No. 20. Test No. 55.	Boiler No. 21. Test No. 56.
Manner of start and stop and kind of run,	Ordinary.	Ordinary.
Duration, ... hrs.	10.5	10.5
Coal consumed, dry (including wood equivalent), ... lbs.	2,842	3,110
Percentage of ash, ... per cent.	14.1	14.5
Water evaporated, ... lbs.	20,649	23,581
Coal per hour, ... lbs.	270.8	296.4
Coal per hour per square foot of grate, . lbs.	10.46	11.46
Water per hour, ... lbs.	1,967.7	2,247.3
Water per hour per square foot of heating surface, ... lbs.	2.1	2.3
Horse-power developed, . . . H. P.	68.4	78.7
Boiler pressure, ... lbs.	56.3	63
Temperature of feed-water, . . . deg.	39.5	39.5
Temperature of escaping gases, . . deg.	467	474
Draught suction, ... in.	0.37	0.39
Water per pound of coal, ... lbs.	7.24	7.58
Water per pound of coal from and at 212 degrees, ... lbs.	8.76	9.17
Water per pound of combustible from and at 212 degrees, ... lbs.	10.23	10.74

NOTE. — The coal when fired contained 3 1-2 per cent. of moisture.

The tests on Boilers No. 20 and No. 21 were made to determine the economy of admitting air to the furnace over the fuel, when using a mixture of pea and dust and Cumberland coal. The two tests were conducted simultaneously with fuel which had been thoroughly mixed so as to obtain the same quality for both boilers. The evaporative results show that the use of air above the fuel, in Boiler No. 20, was attended with a loss of 4.5 per cent. based on coal, and 4.8 per cent. based on combustible. Boiler No. 21 was subsequently tried with anthracite broken coal, containing 13.3 per cent. of ash, and burning 10.5 pounds of coal per square foot of grate per hour, with a draught of 0.29 of an inch, and gave an evaporation of

11.04 pounds of water per pound of combustible from and at 212 degrees. This is a favorable result considering the temperature of the gases, which was 463 degrees.

Boiler No. 22.

Kind of boiler,	Horizontal return tubular.
Number used,	Two.
Horse-power (collective, basis 12 square feet),	Two hundred and twenty.
Kind of coal,	Anthracite, Lehigh, egg.
Age,	Twelve years.

Boiler No. 22 is of the ordinary horizontal tubular type, arranged and set in the general manner shown in the cuts of Boiler No. 3. In this case the flue gases pass directly to the chimney, the top of the boiler being covered with brick work, and the top of the bridge wall is curved upward to conform to the curve of the shell. This boiler is fitted with a much smaller grate than is usually employed, and for this reason the proportion of heating surface to grate surface, which is 68 to 1, is large, and that of the grate to tube area is small.

Dimensions of Boiler No. 22.

Diameter of shell,	60	in.
Length between heads and length of tubes,	17	ft.
Number of tubes (collective) 3 inches outside diameter,	192	
Area of heating surface,	2,680	sq. ft.
Area of grate surface,	39	sq. ft.
Area through tubes,	7.9	sq. ft.
Width of air spaces and metal bars in grates,	Air 3-8 in., metal 7-16 in.	
Ratio of heating surface to grate surface,	68	to 1.
Ratio of grate surface to tube opening,	4.9	to 1.

Results of Test, Boiler No. 22.

	Test No. 57.	
Manner of start and stop and kind of run,	Ordinary.	
Duration,	10	hrs.
Coal consumed,	4,508	lbs.
Percentage of ash,	12.9	per cent.
Water evaporated,	41,714	lbs.
Coal per hour,	450.8	lbs.
Coal per hour per square foot of grate,	11.5	lbs.
Water per hour,	4,171.4	lbs.
Water per hour per square foot of heating surface,	1.6	lbs.
Horse-power developed,	124.2	H. P.
Boiler pressure,	75	lbs.

Temperature of feed water, ...	206	deg.
Temperature of escaping gases, ...	350	deg.
Water per pound of coal, ...	9.25	lbs.
Water per pound of coal from and at 212 degrees, .	9.62	lbs.
Water per pound of combustible from and at 212 degrees,	11.03	lbs.

The test on Boiler No. 22 shows the performance of a boiler having a large proportion of heating surface, and using a standard grade of anthracite coal. Although the amount of power developed is much below the nominal capacity, a reasonably high rate of combustion is maintained, and the resulting evaporation of 11.03 pounds of water from and at 212 degrees per pound of combustible, is favorable when compared with the most economical work. This test shows that a boiler which is called upon to do light work can be made to give economical results with a suitable adaptation of grate surface.

Boiler No. 23.

Kind of boiler, ...	Horizontal return tubular.
Number used, ...	Three.
Horse-power (collective, basis 12 square feet),	Four hundred and fifty.
Kind of coal, ...	Anthracite Chestnut No. 2.
Age, ...	Three months.

Boiler No. 23 embraces a plant of three horizontal tubular boilers set in one battery of brick work, the general features of which are shown in longitudinal section in the following cut. The shells are 72 inches in diameter, and the proportion of heating surface to grate surface is much larger than that ordinarily found in small boilers, being 60 to 1. In this respect the proportions are similar to those of the double deck type of boiler. The brick setting is arranged for admitting a supply of air at the bridge wall. Cast iron perforated globes are provided for this purpose in the manner shown in the cut, and these receive air from the outside through a pipe leading to the rear end of the boiler. A steam jet is employed at the end of the pipe to increase the quantity of air otherwise supplied, and the steam mingles with the air thus introduced.

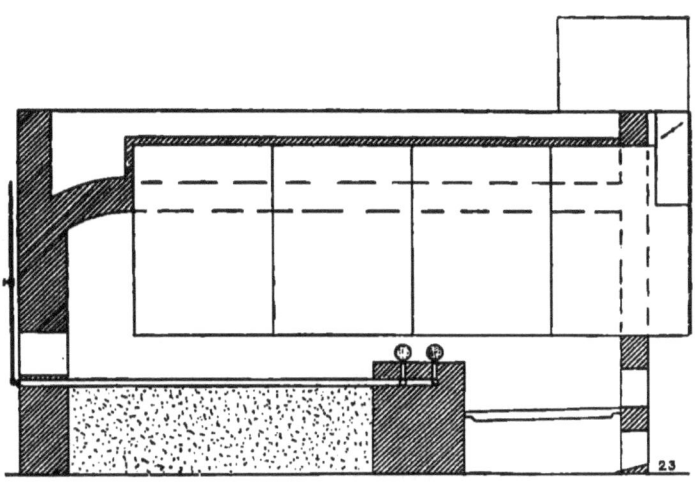

BOILER NO. 23, LONGITUDINAL SECTION.
Dimensions of Boiler No. 23.

Diameter of shell,	72 in.
Length between heads and length of tubes,	16 ft.
Number of tubes (collective) 3 inches outside diameter,	420
Area of heating surface,	5,412 sq. ft.
Area of grate surface,	90 sq. ft.
Area through tubes,	17.3 sq. ft.
Area through flue,	12.6 sq. ft.
Height of chimney,	85 ft.
Width of air spaces and metal bars in grates,	1-2 in.
Distance of grate to shell,	28 in.
Distance of flat bridge to shell,	10 in.
Ratio of heating surface to grate surface,	60 to 1
Ratio of grate surface to tube area,	5.2 to 1
Ratio of grate surface to flue area,	7.1 to 1

Results of Test, Boiler No. 23.

	Test No. 58.
Manner of start and stop and kind of run,	Ordinary.
Duration,	10.7 hrs.
Coal consumed, dry (including wood equivalent),	9,901 lbs.
Percentage of ash,	15.7 per cent.
Coal per hour,	927.9 lbs.
Coal per hour per square foot of grate,	10.3 lbs.
Water per hour,	6,979.8 lbs.
Water per hour per square foot of heating surface,	1.3 lbs.
Horse-power developed,	241.5 H P.

Boiler pressure, ...	80	lbs.
Temperature of feed-water, ...	44	deg.
Temperature of escaping gases, ...	316	deg.
Draught suction, ...	0.32	in.
Water per pound of coal, ...	7.52	lbs.
Water per pound of coal from and at 212 degrees, . .	9.10	lbs.
Water per pound of combustible from and at 212 degrees, .	10.78	lbs.

NOTE.— The coal when fired contained 7 per cent. of moisture.

The test on Boiler No. 23 shows the performance of a plant of large boilers with one of the small grades of anthracite coal. The boiler absorbed nearly the whole of the available heat, and the temperature of the waste gases was reduced below that of the water in the boiler. The evaporative result secured is favorable, if account be taken of the kind of coal and the moist condition in which it was fired. The quantity of power developed is but little over one-half of the rated power of the boilers. Capacity here is sacrificed for economy. It is questionable whether the sacrifice is a wise one from a commercial point of view, for nearly, if not quite, as good economy can be obtained from boilers of smaller diameter, containing a less number of tubes, and consequently a smaller amount of heating surface.

Boiler No. 24.

Kind of boiler, ...	Horizontal return tubular.
Number used, ...	One.
Horse-power (basis 12 square feet), . .	Fifty-three.
Kind of coal, ...	Anthracite broken.
Age, ...	Several years.

Boiler No. 24 is an ordinary horizontal tubular boiler, arranged and set in the same manner as the return tubular boiler shown in the cuts of Boiler No. 1, with the exception that the boiler under consideration has an overhanging front. In this boiler the ratio of heating surface to grate surface is too small to conform to the best practice. The boiler is set only 17 inches from the grate, and the space over the bridge wall is contracted to a height of 4 inches.

Dimensions of Boiler No. 24.

Diameter of shell,	48 in.
Length between heads and length of tubes,	15 ft.
Number of tubes 3 inches outside diameter,	49
Area of heating surface,	639 sq. ft.
Area of grate surface,	24 sq. ft.
Area through tubes,	2 sq. ft.
Area through flue,	1.3 sq. ft.
Height of chimney,	55 ft.
Width of air spaces and metal bars in grates,	Air 5-8 in., metal 7-8 in.
Distance of grate to shell,	17 in.
Distance of bridge wall to shell,	4 in.
Ratio of heating surface to grate surface,	26.6 to 1
Ratio of grate to tube area,	11.9 to 1

Results of Test, Boiler No. 24.

Test No. 59.

Manner of start and stop and kind of run,	Ordinary.	
Duration,	11.2	hrs.
Coal consumed (including wood equivalent),	2,138	lbs.
Percentage of ash,	12	per cent.
Water evaporated,	18,224	lbs.
Coal per hour,	190	lbs.
Coal per hour per square foot of grate,	7.9	lbs.
Water per hour,	1,619.9	lbs.
Water per hour per square foot of heating surface,	2.5	lbs.
Horse-power developed,	48.1	H. P.
Boiler pressure,	59	lbs.
Temperature of feed-water,	210.5	deg.
Temperature of escaping gases,	474	deg.
Draught suction,	0.30	in.
Water per pound of coal,	8.52	lbs.
Water per pound of coal from and at 212 degrees,	8.78	lbs.
Water per pound of combustible from and at 212 degrees,	9.87	lbs.

The test on Boiler No. 24 exhibits the performance of a boiler working under unfavorable conditions and giving a low economic result. Here the low rate of combustion, the high temperature of the gases, and a somewhat large percentage of ash, indicate the main cause of the low degree of economy. The large amount of draught suction in the main flue, which was 0.3 inches, is noticeable in view of the fact that the boiler worked to less than its normal capacity. The reason for this is seen in the small flue area, which was only one-eighteenth of the grate surface.

Boiler No. 25.

Kind of boiler,	Horizontal return tubular.
Number used,	One.
Horse-power (basis 12 square feet),	Eighty-eight.
Kind of coal,	Anthracite White Ash broken.
Age,	Six months.

Boiler No. 25 consists of an ordinary horizontal tubular boiler arranged in the general manner shown in the cut of Boiler No. 10. The flue gases, instead of passing over the top of the shell, as there represented, proceed directly from the smoke arch to the chimney, and the top of the shell is covered with brick. This boiler was one of a battery of two, and the only one in use.

Dimensions of Boiler No. 25.

Diameter of shell,	60	in.
Length between heads and length of tubes,	16	ft.
Number of tubes 3 inches outside diameter,	80	
Area of heating surface,	1,047	sq. ft.
Area of grate surface,	23.7	sq. ft.
Area through tubes,	3.3	sq. ft.
Area through flue,	4	sq. ft.
Height of chimney,	62	ft.
Width of air spaces and metal bars in grates,	Air 5-8 in., metal 3-8 in.	
Distance of grate to shell,	18	in.
Distance of bridge to shell,	7	in.
Ratio of heating surface to grate surface,	44.2 to 1.	
Ratio of grate surface to tube area,	7.1 to 1.	

Results of Test, Boiler No. 25.

Test No. 60.

Manner of start and stop and kind of run,	Ordinary.	
Duration,	10.5	hrs.
Coal consumed (including wood equivalent),	2,351	lbs.
Percentage of ash,	15.7	per cent.
Water evaporated,	20,219	lbs.
Coal per hour,	223.9	lbs.
Coal per hour per square foot of grate,	9.4	lbs.
Water per hour,	1,925.6	lbs.
Water per hour per square foot of heating surface,	1.8	lbs.
Horse-power developed,	58.4	H. P.
Boiler pressure,	60	lbs.
Temperature of feed-water,	202	deg.
Temperature of escaping gases,	369	deg.
Draught suction,	0.11	in.

Water per pound of coal, ... 8.60 lbs.
Water per pound of coal from and at 212 degrees, . 8.94 lbs.
Water per pound of combustible from and at 212 degrees, . 10.61 lbs.

The test on Boiler No. 25 shows the performance of a well proportioned boiler working under favorable conditions for securing economy. There is an ample amount of heating surface; the quantity of power developed, although small compared with the nominal capacity of the boiler, is sufficient to secure a moderately high rate of combustion, and the temperature of the waste gases does not show much loss of heat to the chimney; yet the evaporation per pound of combustible from and at 212 degrees, with a standard grade of anthracite coal, is only 10.61 pounds. The only reason which can be assigned for the unfavorable character of this result appears to be an inferior quality of fuel, which may be inferred from the large percentage of ash, this being 15.7 per cent.

Boilers No. 26 and No. 27.

Kind of boilers, ... Horizontal return tubular.
Number used, each, ... Two.
Horse-power (collective), each (basis 12 square feet), ... Two hundred and eighty.
Kind of coal, ... Nova Scotia Culm.
Age, each, ... Three months.

Boilers No. 26 and No. 27 are of the ordinary horizontal tubular type, and each embraces a plant of two boilers set in one battery of brick work. They are identical with respect to type, arrangement of setting, and location with reference to the chimney, in all respects except one: the side walls of the furnaces and the top of the bridge walls in Boiler No. 27 are provided with perforated tiles through which air is admitted to the furnaces above the fuel, while Boiler No. 26 has no such provision. The two batteries of boilers are located at the same distance from the chimney, one set being on one side of the chimney, and one set on the other side, and a similar plan is followed in each case in the arrangement of the flues. The arrangement of the setting of the boilers is the same in general features as that shown in the longitudinal section of Boiler

No. 6. These boilers are different, however, in respect to the arrangement of smoke arches from those shown, being provided with flush fronts. One boiler in each battery has 100 3½ inch tubes, and the other 140 3-inch tubes. The total area of opening into the air ducts of each battery is 32 square inches. The area of opening through the registers of the fire doors amounts to 22 square inches for each battery.

Dimensions of Boilers No. 26 and No. 27.

Diameter of shell, each,	72 in.
Length between heads and length of tubes,	16 ft.
Number of tubes 3 inches outside diameter, each,	140
Number of tubes 3 1-2 inches outside diameter, each,	100
Area of heating surface, each,	3,347 sq. ft.
Area of grate surface, each,	84 sq. ft.
Area through tubes, each,	11.52 sq. ft.
Width of air spaces and metal bars in grates, each,	3-8 in.
Distance of grate to shell, each,	27 in.
Distance of flat bridge to shell, each,	9 in.
Ratio of heating surface to grate surface, each,	39.8 to 1
Ratio of grate to tube area, each,	7.3 to 1

Results of Tests. Boilers No. 26 and No. 27. (Average of two days).

	Test No. 61. Boiler No. 26.	Test No. 62. Boiler No. 27
Manner of start and stop and kind of run,	Ordinary.	Ordinary.
Duration, ... hrs.	10.5	10.5
Coal consumed, dry (including wood equivalent), ... lbs.	12,714	12,472
Percentage of ash, ... per cent.	9.4	9.7
Water evaporated, ... lbs.	91,653	90,894
Coal per hour, ... lbs.	1,210.9	1,187.8
Coal per hour per square foot of grate, lbs.	14.4	14.1
Water per hour, ... lbs.	8,728.9	8,656.6
Water per hour per square foot of heating surface, ... lbs.	2.6	2.6
Horse-power developed, . . . H. P.	298.8	296.3
Boiler pressure, ... lbs.	79	79
Temperature of feed-water, . . deg.	57	57
Temperature of escaping gases, . . deg.	528	513
Draught suction, ... in.	0.35	0.35
Water per pound of coal, . . . lbs.	7.21	7.28
Water per pound of coal from and at 212 degrees, ... lbs.	8.65	8.74
Water per pound of combustible from and at 212 degrees, ... lbs.	9.51	9.65

The tests on Boilers No. 26 and No. 27 had for an object the determination of the economy produced by admitting air above the fuel, so far as could be determined by tests made on two different sets of boilers. The tests were conducted simultaneously and the coal was used from the same pile, previously mixed to insure uniformity. The registers in the fire-doors were open in both sets of boilers.

The results of the tests show that the admission of air through the passages in the walls increased the evaporation per pound of coal one per cent., and per pound of combustible $1\frac{1}{2}$ per cent. This occurred with Nova Scotia coal, which is of an extremely gaseous and smoky character.

The results of the tests as a whole show a rather low economy when compared, for example, with the work of Cumberland coal; but, as a partial explanation, it is seen that the rate of combustion is large and the temperature of the escaping gases is somewhat excessive.

Boilers No. 28 and No. 29.

Kind of boilers,	Horizontal return tubular.
Number used, each,	One.
Horse-power, each (basis 12 sq. ft.),	No. 28, 150; No. 29, 130.
Kind of coal,	Nova Scotia Culm.
Age,	Three months.

Boilers No. 28 and No. 29 are the two horizontal tubular boilers previously referred to as Boiler No. 26. They are set in one battery of brick work, and the style of the setting is that shown in the cut of Boiler No. 6, with the exception that no provision is made for admitting air above the fuel, and the boilers are provided with flush fronts. The two boilers are duplicates except as to size and number of tubes. In Boiler No. 28 there are 100 3-inch tubes, and in Boiler No. 29 there are 100 $3\frac{1}{2}$ inch tubes.

Dimensions, Boilers No. 28 and No. 29.

Diameter of shell, each,		72 in.
Length of shell between heads and length of tubes, each,		16 ft.
	Boiler No. 28.	Boiler No. 29.
Number of tubes,	140	100
Outside diameter of tubes,	3 in.	3.5 in.
Area of heating surface,	1,799 sq. ft.	1,548 sq. ft.

Area of grate surface,	42 sq. ft.	42 sq. ft.
Area through tubes,	5.8 sq. ft.	5.8 sq. ft.
Ratio of heating surface to grate surface,	42.8 to 1	36.8 to 1
Ratio of grate surface to tube area,	7.3 to 1	7.3 to 1
Width of air spaces and metal bars in grates, each,		3-8 in.
Distance of grate to shell, each,		27 in.
Distance of flat bridge to shell, each,		9 in.

Results of Tests, Boilers No. 28 and No. 29.

	Test No. 63. Boiler 28.	Test No. 64. Boiler 29.
Manner of start and stop and kind of run,	Ordinary.	Ordinary.
Duration, ... hrs.	10.5	10.5
Coal consumed, dry (including wood equivalent), ... lbs.	5,562	5,518
Percentage of ash, ... per cent.	8.5	9.6
Water evaporated, ... lbs.	41,660	39,537
Coal per hour, ... lbs.	529.7	525.5
Coal per hour per square foot of grate, lbs.	12.6	12.5
Water per hour, ... lbs.	3,967.6	3,765.4
Water per hour per square foot of heating surface, ... lbs.	2.2	2.4
Horse-power developed, . . . H. P.	135.8	129
Boiler pressure, ... lbs.	81	81
Temperature of feed-water, . . deg.	57	56
Temperature of escaping gases, . . deg.	489	480
Draught suction, ... in.	0.26	0.25
Water per pound of coal, . . . lbs.	7.49	7.17
Water per pound of coal from and at 212 degrees, ... lbs.	8.99	8.60
Water per pound of combustible from and at 212 degrees, ... lbs.	9.79	9.49

The tests on Boilers No. 28 and No. 29 had for an object the determination of the effect which the size of tubes has upon the economy. This object cannot fairly be said to have been attained, because the boiler which has the small tubes has such a large number of them that the heating surface is increased a considerable amount above that of the other boiler. The test is therefore a determination of the effect of increased heating surface as well as of reduced diameter of tubes. Boiler No. 28, with the small tubes, gave the better result, the increased evaporation being 4.5 per cent. based on coal, and 3.2 per cent. based on combustible.

It is to be noted that the temperature of the escaping gases, here given, is highest in case of the boiler with the larger area of heating surface. It is doubtful whether these figures indicate the true temperature. The temperature was taken in each case at a point in the main flue near the smoke arch, and as the flue was common to both boilers, and both boilers were in operation while the test was going on, it is probable that the temperature given is in some degree an average for both boilers, rather than the actual temperature for the single boiler tested.

Boiler No. 30.

Kind of boiler,	Horizontal return tubular.
Number used,	One.
Horse-power (basis 12 square feet),	Fifty-three.
Kind of coal,	Cumberland bituminous.
Age,	Several years.

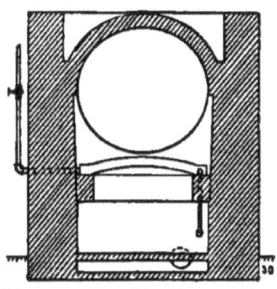

BOILER No. 30, CROSS SECTION THROUGH FURNACE.

Boiler No. 30 is an ordinary horizontal tubular boiler having a special arrangement of furnace for burning bituminous coal. The general features of the furnace and its appliances are shown in the following cuts. Several jets of superheated steam are introduced beneath the grate, and a supply of air is forced in above the fuel by means of a blower. The steam is derived from the boiler, and it receives its superheat from a cast iron heater placed at the back end of the furnace. The quantity of steam used amounted to about 8 per cent. of the whole, and this has been deducted from the total evaporation to determine the net quantities given in the table of results.

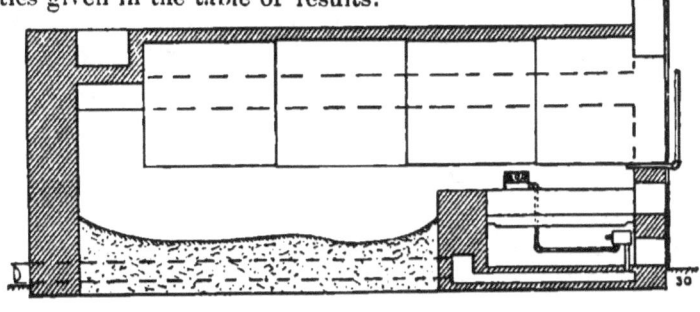

BOILER No. 30, LONGITUDINAL SECTION.

BOILER No. 30.

Dimensions of Boiler No. 30.

Diameter of shell,	48	in.
Length of shell between heads and length of tubes,	15	ft.
Number of tubes 3 inches outside diameter,	49	
Area of heating surface,	640	sq. ft.
Area of grate surface,	13.5	sq. ft.
Area through tubes,	2	sq. ft.
Area through flue,	2	sq. ft.
Distance of grate to shell,	19	in.
Distance of flat bridge to shell,	6	in.
Ratio of heating surface to grate surface,	47	to 1
Ratio of grate to tube area,	6.6	to 1

Results of Tests, Boiler No. 30.

	Test No. 65.	
Manner of start and stop,	Thin fire.	
Kind of run,	Continuous.	
Duration,	7.6	hrs.
Coal consumed, dry,	1,361	lbs.
Percentage of ash,	6.5	per cent.
Water evaporated,	10,442	lbs.
Coal per hour,	177.9	lbs.
Coal per hour per square foot of grate,	13.2	lbs.
Water per hour,	1,365	lbs.
Water per hour per square foot of heating surface,	2.1	lbs.
Horse-power developed,	47.2	H. P.
Boiler pressure,	75	lbs.
Temperature of feed water,	44	deg.
Temperature of escaping gases,	362	deg.
Water per pound of coal,	7.67	lbs.
Water per pound of coal from and at 212 degrees,	9.28	lbs.
Water per pound of combustible from and at 212 degrees,	9.91	lbs.

The test on Boiler No. 30 had for an object the determination of the general performance of the system of combustion here employed. This system produced an almost smokeless furnace. A very small quantity of light smoke appeared for a short time after firing fresh coal, but it was almost colorless. The high character of combustion, which the absence of smoke seemed to indicate, secured no apparent benefit in the matter of economy, for after allowing for the steam used by the apparatus, the result is only 9.91 pounds of water from and at 212 degrees per pound of combustible, which is some 20 per cent. below the best practice. This inferior performance cannot be attributed to any unfavorable conditions which the test shows regarding the quality of fuel, rate of combustion, or the temperature of the flue gases, for these are such as ordinarily give good results.

Boiler No. 31.

Kind of boiler,	Horizontal return tubular.
Number used,	Three.
Horse power (collective basis 12 square feet),	Three hundred and twelve.
Age,	One month.

Boiler No. 31 consists of a plant of three horizontal tubular boilers, set in one battery of brick work. The general features of the setting are similar to those shown in the cut of Boiler No. 5, with the exception that the bridge walls are arched upward to conform to the curve of the shell. These boilers are deficient in flue area, this being about one third of the area for draught through the tubes. As a consequence, a strong draught is required in the flue to secure a relatively slow rate of combustion and small capacity.

Dimensions of Boiler No. 31.

Diameter of shell,	66 in.
Length between heads and length of tubes,	15 ft.
Number of tubes (collective) 3 inches outside diameter,	306
Area of heating surface,	3,374 sq. ft.
Area of grate surface,	90 sq. ft.
Area through tubes,	12.6 sq. ft.
Area through flue,	4.6 sq. ft.
Height of chimney,	75 ft.
Width of air spaces and metal bars in grates,	1-2 in.
Distance of grate to shell,	24 in.
Distance of curved bridge to shell,	12 in.
Ratio of heating surface to grate surface,	41.6 to 1
Ratio of grate surface to tube area,	7.1 to 1
Ratio of grate surface to flue area,	18.3 to 1

Results of Tests, Boiler No. 31.

	Test No. 66.	Test No. 67.
Kind of coal,	George's Creek Cumberland.	Coke from gas coal.
Manner of start and stop,	Thin fire.	Thin fire.
Kind of run,	Factory.	Factory.
Duration, ... hrs.	11.7	11.7
Coal consumed, dry, . . . lbs.	7,330	8,541
Percentage of ash, . . per cent.	6.6	4.9
Water evaporated, . . . lbs.	75,119	79,371
Coal per hour . . . . lbs.	626.5	730
Coal per hour per square foot of grate, lbs.	7	8.1
Water per hour, ... lbs.	6,820.4	6,783.8
Water per hour per square foot of heating surface, ... lbs.	1.8	1.8
Horse power developed, . . H. P.	204.1	202.9
Boiler pressure, ... lbs.	84	88
Temperature of feed-water, . deg.	211	211
Temperature of escaping gases, . deg.	431	428
Draught suction, ... in.	0.31	0.26
Water per pound of coal, . . lbs.	10.89	9.29
Water per pound of coal from and at 212 degrees, ... lbs.	11.33	9.66
Water per pound of combustible from and at 212 degrees, ... lbs.	12.07	10.11

The tests on Boiler No. 31 had for an object the determination of its general economy, and the relative economy produced with Cumberland coal and gas house coke. The evaporative result obtained with Cumberland coal on Test No. 66 compares favorably with the best work. Although the boilers operated under a low rate of combustion and developed considerably less than their nominal capacity, the flue gases passed off at a somewhat high temperature. This may be explained by the fact of the new condition of the settings which prevented undue loss from the admission of superfluous air through the brick work. Comparing the results obtained with the two different fuels, the water evaporated per pound of coke is 14.7 per cent. less than that with the standard grade of coal. On the basis of prices which existed at the time of the tests, coal being $3.25 per ton and coke $3.00 per ton (2000 pounds), the cost of coal required to produce a given amount of steam is 7.4 per cent. less than the cost of coke.

Boiler No. 32.

Kind of boiler,	Horizontal return tubular.
Number used,	Two.
Horse-power (collective, basis 12 sq. ft.),	One hundred and seventy-five.
Kind of coal,	George's Creek Cumberland.
Age,	Ten years.

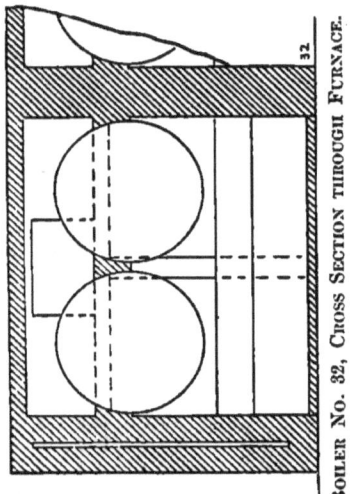

Boiler No. 32 embraces a plant of two horizontal tubular boilers, arranged in the manner shown in the following cuts. The boilers are set over a single furnace, and the upper surfaces of the shells are exposed to the heat of the escaping gases on their way from the front smoke arch to the flue, thereby furnishing a small amount of steam heating surface. The boiler had been in service for several years, but the inside surfaces were free from scale, and both boiler and settings were in every way in good condition. This boiler was the end boiler of a large plant, and the boiler next to it was in operation during the progress of the tests.

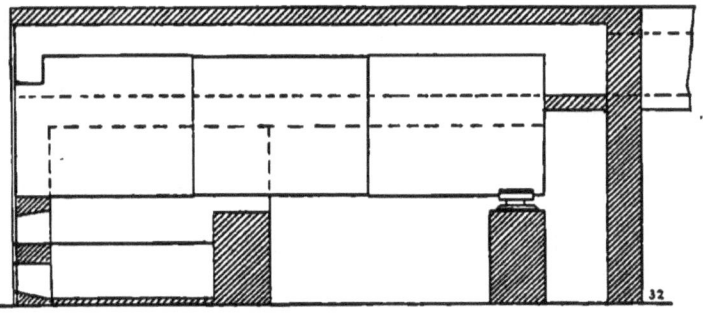

BOILER NO. 32, LONGITUDINAL SECTION.

BOILER No. 32.

Dimensions of Boiler No. 32.

Diameter of shell,	54 in.
Length between heads and length of tubes,	15 ft.
Number of tubes (collective) 3 inches outside diameter,	160
Area of water-heating surface,	1,980 sq. ft.
Area of steam-heating surface,	120 sq. ft.
Area of total heating surface,	2,100 sq. ft.
Area of grate surface,	49.3 sq. ft.
Area through tubes,	6.6 sq. ft.
Area through flue,	6 sq. ft.
Width of air spaces and metal bars in grates,	Air 7-16 in., metal 5-16 in.
Distance of grate to shell,	19 in.
Distance of flat bridge wall to shell,	5 in.
Ratio of water-heating surface to grate surface,	40 to 1
Ratio of grate surface to tube area,	7.5 to 1

Results of Tests, Boiler No. 32. (Average of two).

	Test No. 68.
Manner of start and stop and kind of run,	Ordinary.
Duration,	11.5 hrs.
Coal consumed, dry (including wood equivalent),	6,288 lbs.
Percentage of ash,	6.5 per cent.
Water evaporated,	58,993 lbs.
Coal per hour,	546.8 lbs.
Coal per hour per square foot of grate,	11.1 lbs.
Water per hour,	5,134.2 lbs.
Water per hour per square foot of water-heating surface,	2.6 lbs.
Horse-power developed,	177.6 H. P.
Boiler pressure,	65 lbs.
Temperature of feed-water,	57 deg.
Temperature of escaping gases,	408 deg.
Draught suction,	0.35 in.
Percentage of moisture in steam,	2.2 per cent.
Water per pound of coal,	9.39 lbs.
Water per pound of coal from and at 212 degrees,	11.20 lbs.
Water per pound of combustible from and at 212 degrees,	11.98 lbs.

The test on Boiler No. 32 shows the performance of an ordinary tubular boiler worked with Cumberland coal. The percentage of ash is small and indicates a good quality of fuel. The rate of combustion is moderately high, and the temperature of the escaping gases is not excessive for bituminous coal. These conditions are favorable for economy, and the resulting evaporation bears out the expectations which they justify. The effect of the steam heating surface in this boiler does not appear to have been sufficient to thoroughly dry the steam.

The calorimeter test showed that the steam contained 2.2 per cent. of moisture. The location of this surface is favorable for the deposit of soot and ashes, and it is thereby rendered inefficient for its purpose.

Boiler No. 33.

Kind of boiler,	Horizontal return tubular.
Number used,	Two.
Horse-power (collective, basis 12 sq. ft.),	One hundred and fifty-eight.
Kind of coal,	Two parts Anthracite Lehigh Buckwheat, one part Clearfield Bituminous.
Age,	Five years.

Boiler No. 33 consists of a plant of two horizontal tubular boilers set in one battery of brick work. The general features of the boiler and the style of setting are similar to those shown in the cut of Boiler No. 6. The boilers are fitted with pipes placed beneath the shell near the side walls, and the feed water is first passed through these pipes. The additional surface thus exposed amounted to 180 square feet, and this is included in the quantity given in the table of dimensions. The side walls and top of the bridge wall, as shown in the cut referred to, are provided with perforated tiles for the admission of air above the fuel, and the furnaces are fitted with fire doors of special form, through which a large amount of air is also admitted. The plant is provided with a flue heater consisting of vertical cast iron pipes, arranged in sections and connected together by means of two headers placed outside the brick work, one at the lower end and one at the upper end. The water supplied to the heater enters it through the lower header and leaves it through the upper header. The exterior surfaces of the pipes are kept clean by means of scrapers, worked alternately up and down, and operated by power. This is placed in a direct line between the boilers and the chimney, as indicated in the ground plan given in the following cut. The flue, which is provided for carrying the gases directly to the chimney, lies beneath the chamber which encloses the heater. The area of surface exposed to the heat in this apparatus is nearly as large as the total area of heating surface of the two boilers. The third boiler shown in the plan was out of use during the tests.

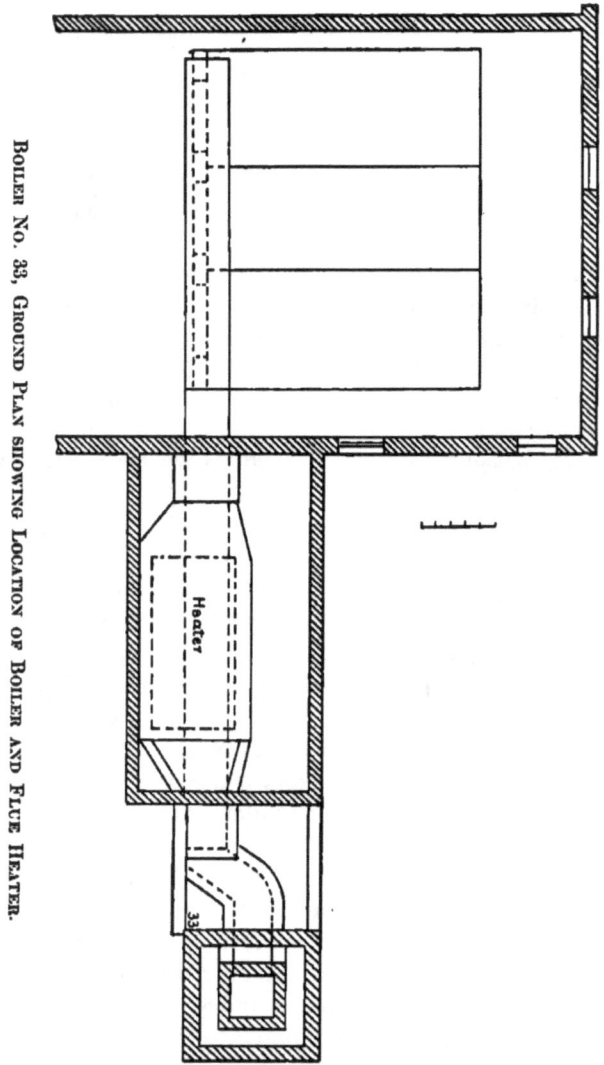

BOILER NO. 33, GROUND PLAN SHOWING LOCATION OF BOILER AND FLUE HEATER.

Dimensions of Boiler No. 33.

Diameter of shell,	60 in.
Length between heads and length of tubes,	15 ft.
Number of tubes (collective) 3 inches outside diameter,	132
Area of heating surface,	1,894 sq. ft.
Area of grate surface,	50 sq. ft.
Area through tubes,	5.4 sq. ft.
Height of chimney,	110 sq. ft.
Width of air spaces and metal bars in grates,	Air 3-8 in., metal 7-16 in.
Ratio of heating surface to grate surface,	37.9 to 1
Ratio of grate surface to tube opening,	9.3 to 1
Area of heating surface in flue heater,	1,600 sq. ft.

Results of Tests, Boiler No. 33.

	Test No. 60.	Test No. 70.
	Heater in use.	Heater not in use.
Manner of start and stop and kind of run,	Ordinary.	Ordinary.
Duration, hrs.	12	12
Coal consumed, dry (including wood equivalent), lbs.	5,901	6,564
Percentage of ash, per cent.	11.8	11.8
Water evaporated, lbs.	51,955	52,303
Coal per hour, lbs.	491.7	547
Coal per hour per square foot of grate, lbs.	9.8	10.9
Water per hour, lbs.	4,329.5	4,358.5
Water per hour per square foot of heating surface, lbs.	2.3	2.3
Horse-power developed, H. P.	145	146.3
Boiler pressure, lbs.	77	77
Temperature of feed-water entering heater, deg.	95	–
Temperature of feed-water entering boiler, deg.	175	93
Temperature of escaping gases leaving boiler, deg.	346	390
Temperature of escaping gases leaving heater, deg.	231	–
Water per pound of coal, lbs.	8.80	7.97
Water per pound of coal from and at 212 degrees, lbs.		9.22
Water per pound of combustible from and at 212 degrees, lbs.		10.45

NOTE.— The coal when fired contained 2 1-2 per cent. of moisture.

The tests on Boiler No. 33 were made to determine the economy produced by a flue heater attached to ordinary horizontal tubular boilers. They have a special interest in view of the fact that the temperature of the escaping gases was not above the point which is ordinarily considered favorable to good results. The use of the heater was attended by a reduction of 115 degrees in the temperature of the gases, these entering the heater at 346 degrees, and leaving it at .231 degrees. The temperature of the water was raised from 95 to 175, or 80 degrees. Comparing the two evaporative results, the water per pound of coal was increased from 7.97 to 8.80 pounds, or 10.5 per cent. The economic result produced when the heater was not in use, compares favorably with that obtained in good practice, considering the inferior grade of coal that was employed.

Boiler No. 34.

Kind of boiler,	Horizontal return tubular.
Number used,	One.
Horse-power (basis 12 square feet),	Sixty-five.
Kind of coal,	Bituminous Walston.
Age,	Four years.

Boiler No. 34 is an ordinary horizontal tubular boiler, arranged and set in the general manner shown in the cut of Boiler No. 5.

Dimensions of Boiler No. 34.

Diameter of shell,	54	in.
Length between heads and length of tubes,	12	ft.
Number of tubes three inches outside diameter,	71	
Area of heating surface,	783	sq. ft.
Area of grate surface,	24	sq. ft.
Area through tubes,	2.9	sq. ft.
Area through flue,	3.1	sq. ft.
Height of chimney,	50	ft.
Width of air spaces and metal bars in grates,	Air 5-16 in., metal 1 1-8 in.	
Distance of grate to shell,	18	in.
Distance of flat bridge to shell,	7	in.
Ratio of heating surface to grate surface,	32.2 to 1	
Ratio of grate surface to tube area,	8.3 to 1	

Results of Test, Boiler No. 34.

Test No. 71.

Manner of start and stop,	Ordinary, with preliminary heating.
Kind of run,	Continuous.
Duration,	8.6 hrs.
Coal consumed, dry (including wood equivalent),	2,562 lbs.
Percentage of ash,	7.3 per cent.

Water evaporated,	20,395	lbs.
Coal per hour,	296	lbs.
Coal per hour per square foot of grate,	12.3	lbs.
Water per hour,	2,357.7	lbs.
Water per hour per square foot of heating surface,	3	lbs.
Horse-power developed,	79.7	H. P.
Boiler pressure,	77	lbs.
Temperature of feed-water,	86	deg.
Temperature of escaping gases,	572	deg.
Water per pound of coal,	7.69	lbs.
Water per pound of coal from and at 212 degrees,	9.27	lbs.
Water per pound of combustible from and at 212 degrees	10.00	lbs.

The test on Boiler No. 34 shows the performance of a horizontal tubular boiler, using Wallston bituminous coal. This fuel is a "free-burning" coal, similar in general characteristics to that known as Pittsburg coal. Judging from the high temperature of the escaping gases, which was 572 degrees, the area of the heating surface does not seem to have been sufficient to attain the best results from this class of fuel. If the loss from the high temperature be measured by the effect which a flue heater would produce on the economy under these circumstances, this may be assumed to represent a loss of at least 10 per cent. The performance with Wallston coal, under favorable conditions, would thus be an evaporation of about 11 pounds of water from and at 212 degrees per pound of combustible. This is about 10 per cent. below the results of good practice for Cumberland bituminous coal.

Boiler No. 35.

Kind of boiler,	Horizontal return tubular.
Number used,	Three.
Horse-power (collective basis 12 sq. ft.),	Two hundred and seventy-five.
Age,	Six years.

Boiler No. 35 embraces a plant of three horizontal tubular boilers, set in one battery of brick work, the general features of which are shown in longitudinal section in the following cut.

BOILER No. 35.

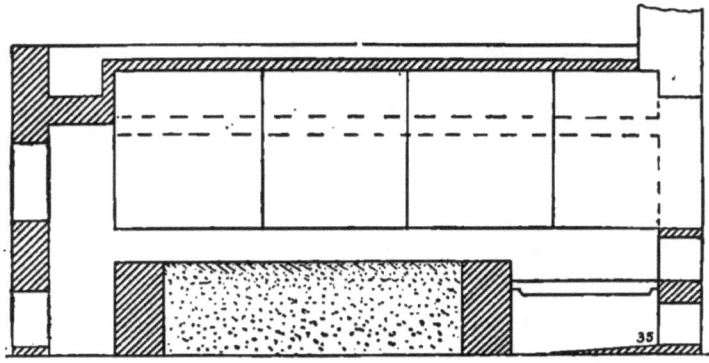

BOILER NO. 35, LONGITUDINAL SECTION.

Dimensions of Boiler No. 35.

Diameter of shell,	60	in.
Length between heads and length of tubes,	17	ft.
Number of tubes (collective,) three inches outside diameter,	231	
Area of heating surface,	3,306	sq. ft.
Area of grate surface,	69.8	sq. ft.
Area through tubes,	9.5	sq. ft.
Area through flue,	10.1	sq. ft.
Height of chimney,	98	ft.
Width of air spaces and metal bars in grates,	3-8	in.
Distance of grate to shell,	19	in.
Distance of flat bridge wall to shell,	12	in.
Ratio of heating surface to grate surface,	47.4 to 1	
Ratio of grate surface to tube area,	7.3 to 1	

Results of Tests, Boiler No. 35.

	Test No. 72.	Test No. 73.
Kind of coal,	George's Creek Cumberland.	Philadelphia and Reading Anthracite broken.
Manner of start and stop and kind of run,	Ordinary.	Ordinary.
Duration, ... hrs.	12	11.7
Coal consumed, dry (including wood equivalent), ... lbs.	5,639	9,765
Percentage of ash, ... per cent.	8.3	10.7
Water evaporated, ... lbs.	50,263	79,868
Coal per hour, ... lbs.	469.9	832.5
Coal per hour per square foot of grate, . lbs.	6.7	11.9
Water per hour, ... lbs.	4,204.3	6,811.0
Water per hour per square foot of heating surface, ... lbs.	1.3	2.1
Horse-power developed, . . . H. P.	141.3	228.7
Boiler pressure, ... lbs.	88	86
Temperature of feed-water, . . . deg.	94	96
Temperature of escaping gases, . . deg.	340	428
Draught suction, ... in.	0.09	0.17
Percentage of moisture in steam, per cent.	0.86	0.49
Water per pound of coal, . . . lbs.	8.91	8.18
Water per pound of coal from and at 212 degrees, ... lbs.	10.34	9.46
Water per pound of combustible from and at 212 degrees, ... lbs.	11.24	10.60

NOTE.—The Cumberland coal when fired contained 3 per cent. of moisture.

The tests on Boiler No. 35 were made to determine the general economy of the boiler with Cumberland and anthracite coal. The test with Cumberland coal was made under conditions of a slow rate of combustion, and that with anthracite coal under a much higher rate, so that the relative performance of the two fuels cannot fairly be compared. The boiler has an ample area of heating surface, the proportion of which to grate surface is 47.4 to 1; but this does not appear to have been sufficient to absorb the whole of the available heat which was generated in the case of the test with anthracite coal, the temperature of the escaping gases being 428 degrees. The efficiency of the surfaces had probably become deteriorated by deposits of scale, and this furnishes a reason for the somewhat unfavorable evaporative result which was produced. The result obtained on the test with Cumberland coal is equally unfavorable, but here the rate of combustion was too low to secure the highest degree of economy.

Boiler No. 36.

Kind of boiler,	Horizontal tubular, detatched furnace.
Number used,	One.
Horse-power (basis 12 sq. ft.),	Two hundred and seventy.
Age,	Two months.

Boiler No. 36 is a horizontal tubular boiler arranged with a detached furnace in the general manner shown in longitudinal section in the following cut. The position of the furnace with reference to the boiler is precisely that of the fire-box of a locomotive boiler. After the products of combustion have passed forward through the tubes, they return beneath the shell and enter an underground flue leading to the chimney. Direct radiation of heat from the coal in the furnace to the tube sheet is prevented by means of a brick arch which overhangs the grate. This boiler has a large shell, long tubes, and a large extent of heating surface compared with grate surface.

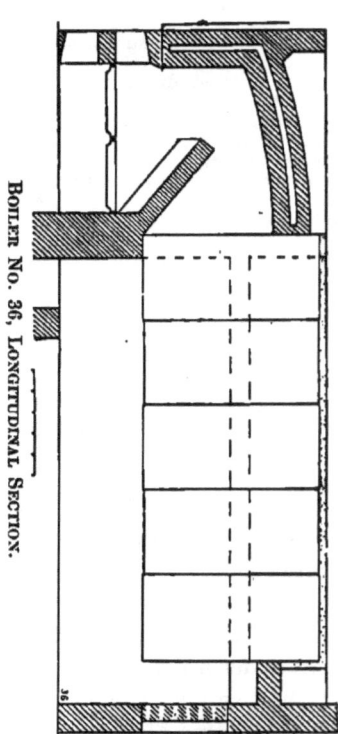

BOILER NO. 36, LONGITUDINAL SECTION.

Dimensions of Boiler No. 36.

Diameter of shell,	96	in.
Length between heads and length of tubes,	20	ft.
Number of tubes 3 inches outside diameter,	206	
Area of heating surface,	3,242	sq. ft.
Area of grate surface,	56	sq. ft.
Area through tubes,	8.5	sq. ft.
Area through flue,	9.7	sq. ft.
Height of chimney,	98	ft.
Width of air spaces and metal bars in grates,	Air 3-8 in., metal 1-2 in.	
Ratio of heating surface to grate surface,	57.9 to 1	
Ratio of grate surface to tube area,	6.7 to 1	

Results of Tests. Boiler No. 36.

	Test No. 74.	Test No. 75.	Test No. 76.	Test No. 77.
Kind of fuel,	Philadelphia and Reading Anthracite Broken.	George's Creek Cumberland.	4 pts. Geo.'s Creek Cumberland, 6 pts. Anthracite Screenings.	Crude Petroleum.
Manner of start and stop and kind of run,	Ordinary.	Ordinary.	Ordinary.	*
Duration, . . . hrs.	12.4	10.8	10.9	4.6
Fuel consumed, dry (including wood equivalent), lbs.	8,242	7,300	7,450	2,327
Percentage of ash, per cent.	10.3	8.3	8.7	–
Water evaporated, lbs.	72,629	69,284	66,671	28,751
Fuel per hour, . lbs.	665.2	675.9	683.5	504.1
Fuel per hour per square foot of grate, . lbs.	11.9	12.1	12.2	–
Water per hour, . lbs.	5,861.5	6,385.4	6,102.2	6,228
Water per hour per square foot of heating surface, lbs.	1.8	2	1.9	1.9
Horse-power developed, H.P.	196.1	214.6	204.8	209.2
Boiler pressure, . lbs.	63	85	85	78
Temperature of feed-water, deg.	94	95	96	93
Temperature of escaping gases, . . deg.	321	397	367	429
Draught suction, . . in.	0.25	0.20	0.31	–
Percentage of moisture in steam, . per cent.	–	0.43	0.49	–
Water per pound of fuel, lbs.	8.81	9.49	8.93	Gross 12.36 / Net 11.80
Water per pound of fuel from and at 212 degrees, lbs.	10.16	10.99	10.33	Net 13.66
Water per pound of combustible from and at 212 degrees, . . lbs.	11.33	11.99	11.30	–

* Afternoon period of the day's run starting with hot furnace and boiler.

NOTE.—The Cumberland coal when fired contained 3 per cent. of moisture and the mixed fuel 4 per cent.

The tests on Boiler No. 36 were made to determine the economy of this particular arrangement of setting, as also the relative economy of four different kinds of fuel, one of which was what is called the "residuum" of crude petroleum. Looking at the individual results of these tests, it appears that the boiler showed a performance with anthracite coal which is seldom exceeded by boilers set in the ordinary manner. The comparative result obtained with Cumberland coal is somewhat less favorable, considering the class of the fuel, though taken by itself it represents excellent work. In view of the new condition of the boiler, however, these tests cannot be held to show special advantage due to the use of a detached furnace. Comparing the results of the coal tests with each other, the evaporation per pound of Cumberland coal is 8.1 per cent. greater than that per pound of anthracite coal; and the evaporation per pound of the mixture is 1 per cent. greater. Basing the comparison on the cost of fuel, the cost of coal required to produce a given amount of steam is 8 per cent. less when Cumberland coal is used, and 23.5 per cent. less when mixed fuel is used, than that of anthracite coal. The prices on which these figures are obtained, per ton of 2,240 pounds, are $4.50 each for anthracite and Cumberland, and $2.75 for screenings.

The apparatus used for supplying the petroleum to the furnaces consisted of four injectors, to which the oil was brought by means of a steam pump. The injectors or burners were placed in a horizontal position in the front wall of the furnace, and pointed toward the middle of the arch. They consisted of two wrought iron tubes one within the other. The outer one carried the oil, and the inner one was arranged so as to supply a jet of steam, by means of which the oil was properly distributed as it entered the furnace. The amount of steam used by the pump and jet was 4.5 per cent. of that generated by the boiler. After deducting this from the total evaporation, the net amount of water from and at 212 degrees evaporated per pound of oil was 13.66 pounds. This quantity is 34 per cent. more than the evaporation per pound of anthracite coal, and 24 per cent. more than the evaporation per pound of Cumberland coal. With Cumberland coal at $4.50 per ton (2,240 pounds), the price of oil required to make the cost of fuel for producing a given amount of steam the same in both cases, is 1.8 cents per gallon.

The result obtained with the residuum is probably inferior to that which would be obtained with the crude oil itself. This oil came from Pennsylvania. A test made on another boiler with oil which was obtained from wells in Canada, gave a net evaporation of 15 pounds of water from and at 212 degrees per pound of oil. The boiler used in this case was of the ordinary horizontal tubular type, 6 feet in diameter and 14 feet long. It contained 71 four-inch tubes, and the ratio of heating surface to grate surface was 39 to 1.

Boiler No. 37.

Kind of boiler,	Horizontal return tubular.
Number used,	One.
Horse-power (basis 12 square feet),	Ninety-five.
Kind of coal,	Bituminous Ohio Lump.
Age,	Several years.

Boiler No. 37 is an ordinary horizontal tubular boiler set in brick work in the general manner shown in longitudinal section in the following cut. The fire door admits a large amount of air into the furnaces above the fuel, being specially arranged for this purpose.

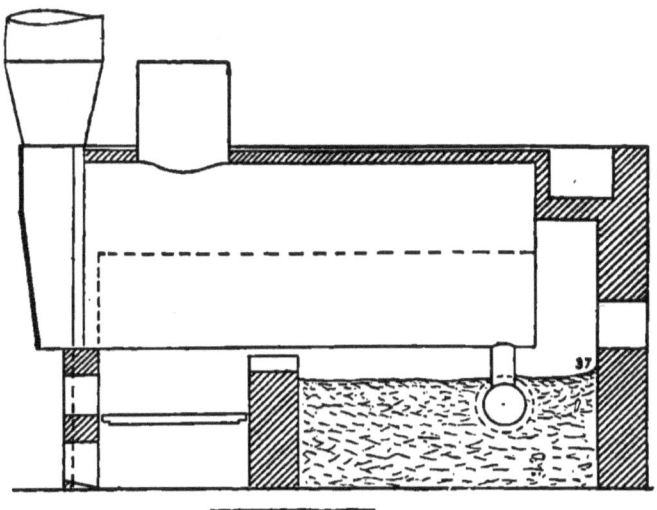

BOILER NO. 37, LONGITUDINAL SECTION.

BOILER No. 37.

Dimensions of Boiler No. 37.

Diameter of shell,	72	in.
Length between heads and length of tubes,	14	ft.
Number of tubes, 4 inches outside diameter,	71	
Area of heating surface,	1,144	sq. ft.
Area of grate surface,	29.2	sq. ft.
Area through tubes,	5.4	sq. ft.
Area through flue,	7	sq. ft.
Height of chimney,	60	ft.
Width of air spaces and metal bars in grates,	3-8	in.
Distance of grate to shell,	24	in.
Distance of curved bridge to shell,	8	in.
Ratio of heating surface to grate surface,	39.2 to 1	
Ratio of grate surface to tube area,	5.4 to 1	

Results of Tests, Boiler No. 37.

Test No. 78.

Manner of start and stop,	\{ Ordinary, with preliminary heating.	
Kind of run,	Continuous.	
Duration,	9.7	hrs.
Coal consumed, dry (including wood equivalent),	3,111	lbs.
Percentage of ash,	7.6	per cent.
Water evaporated,	22,650	lbs.
Coal per hour,	319.1	lbs.
Coal per hour per square foot of grate,	10.9	lbs.
Water per hour,	2,323.1	lbs.
Water per hour per square foot of heating surface,	2	lbs.
Horse-power developed,	79.9	H. P.
Boiler pressure,	74.8	lbs.
Temperature of feed-water,	64.5	deg.
Temperature of escaping gases,	501	deg.
Water per pound of coal,	7.28	lbs.
Water per pound of coal from and at 212 degrees,	8.64	lbs.
Water per pound of combustible from and at 212 degrees,	9.35	lbs.

The test on Boiler No. 37 is of interest in showing the performance of Ohio lump coal in a horizontal tubular boiler. There is a noticeably high temperature of the escaping gases, this being 501 degrees, and it may be concluded that the boiler was deficient in heating surface. If it had been arranged with suitable proportions, it would doubtless have shown an evaporation of 10 pounds of water from and at 212 degrees per pound of combustible, which is $16\frac{2}{3}$ per cent. below the best figures obtained with Cumberland coal. The coal here used, like that on Test No. 71, is what is called "free-burning" coal, and this quality doubtless has something to do with the high flue temperature.

Boiler No. 38.

Kind of boiler,	Horizontal return tubular.
Number used,	One.
Horse-power (basis 12 square feet),	Thirty.
Kind of coal,	Anthracite Wilkesbarre broken.
Age,	One year.

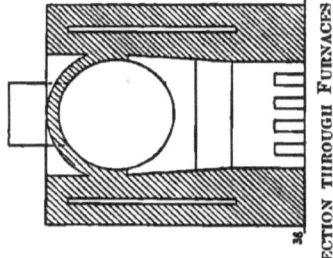

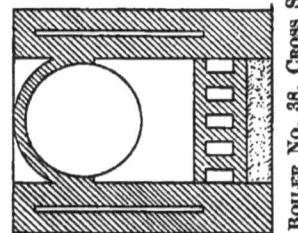

BOILER NO. 38, CROSS SECTION THROUGH FURNACES AND BACK OF BRIDGE WALL.

Boiler No. 38 is a horizontal tubular boiler arranged for superheating the steam, and its general features are shown in the following cuts. The steam-heating surface is here obtained by filling the steam space above the water line with tubes. The products of combustion, after passing through the lower tubes in the usual manner, return through the superheating tubes, and enter the chimney from the rear end. The steam on leaving the boiler enters a vertical drum 30 inches in diameter and 8 feet high. The draft was produced by a blower which discharged under the grates.

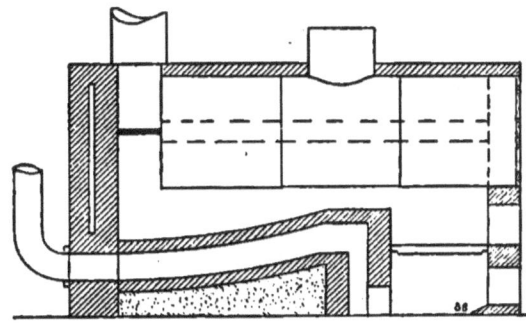

BOILER 38, LONGITUDINAL SECTION.

BOILER No. 38.

Dimensions of Boiler No. 38.

Diameter of shell,	42	in.
Length between heads and length of tubes,	10	ft.
Number of tubes, 2 inches outside diameter, below water-line,	67	
Number of tubes, 2 inches outside diameter, above water-line,	68	
Area of water-heating surface,	369.3	sq. ft.
Area of steam-heating surface,	318.8	sq. ft.
Area of grate surface,	9.2	sq. ft.
Tube area below water-line,	1.1	sq. ft.
Flue area,	2.2	sq. ft.
Height of chimney,	30	ft.
Width of air spaces and metal bars in grates,	5-8	in.
Ratio of water-heating surface to grate surface,	40.3	to 1
Ratio of steam-heating surface to grate surface,	36.6	to 1
Ratio of grate surface to tube area below water-line,	8.2	to 1

Results of Tests, Boiler No. 38.

Test No. 79.

Manner of start and stop,	Ordinary, with preliminary heating.	
Kind of run,	Continuous.	
Duration,	8	hrs.
Coal consumed, (including wood equivalent),	1,453	lbs.
Percentage of ash,	9.5	per cent.
Water evaporated,	10,341	lbs.
Coal per hour,	181.5	lbs.
Coal per hour per square foot of grate,	20	lbs.
Water per hour,	1,291.8	lbs.
Water per hour per square foot of water-heating surface,	3.5	lbs.
Horse-power developed,	45.5	H. P.
Boiler pressure,	66	lbs.
Temperature of feed-water,	36	deg.
Temperature of escaping gases entering upper tubes,	558	deg.
Temperature of escaping gases leaving upper tubes,	394	deg.
Number of degrees of super-heating,	30	deg.
Water per pound of coal,	7.12	lbs.
Water per pound of coal from and at 212 degrees,	8.64	lbs.
Water per pound of combustible from and at 212 degrees,	9.55	lbs.

The test on Boiler No. 38 shows the performance of a super-heating boiler using anthracite coal. The chief interest in the test lies in the effect produced by the steam-heating surface as here arranged. The temperature of the gases in passing over this surface was reduced from 558 degrees, the temperature on entering, to 394 degrees. This served to dry the moisture in the steam and superheat it 30 degrees at the point where it left the drum, and probably 50 degrees at the point of leaving the boiler. The general results of the test are inferior to the best boiler practice. The evaporation per pound of coal, however, is about the same as vertical tubular boilers give, when producing superheated steam at the same temperature.

Boiler No. 39.

Kind of boiler,	Horizontal return tubular.
Number used,	One.
Horse-power (basis 12 square feet),	Twenty-seven.
Kind of coal,	Bituminous Walston.
Age,	Several years.

Boiler No. 39 is an ordinary horizontal boiler set in brick work in the general manner shown in the cut of Boiler No. 10, differing, however, from that boiler in being provided with an overhanging front, the gases from which pass directly to the chimney. This boiler has a small shell, short tubes, and a low proportion of heating surface to grate surface.

Dimensions of Boiler No. 39.

Diameter of shell,	42 in.
Length between heads and length of tubes,	10 ft.
Number of tubes, 3 inches outside diameter,	36
Area of heating surface,	322 sq. ft.
Area of grate surface,	14.4 sq. ft.
Area through tubes,	1.5 sq. ft.
Area through flue,	2.2 sq. ft.
Height of chimney,	56 ft.
Width of air spaces and metal bars in grates,	Air 5-16 in., metal 11-16 in.
Distance of grate to shell,	21 in.
Distance of flat bridge to shell,	8 in.
Ratio of heating surface to grate surface,	22.3 to 1
Ratio of grate surface to tube area,	9.8 to 1

Results of Test, Boiler No. 39.

Test No. 80.

Manner of start and stop,	Ordinary, with preliminary heating.
Kind of run,	Continuous.
Duration,	8.3 hrs.
Coal consumed, dry (including wood equivalent),	1,134 lbs.
Percentage of ash,	7.6 per cent.
Water evaporated,	9,519 lbs.
Coal per hour,	137.4 lbs.
Coal per hour per square foot of grate,	9.52 lbs.
Water per hour,	1,153.3 lbs.

Water per hour per square foot of heating surface,	3.6	lbs.
Horse-power developed,	34.9	H. P.
Boiler pressure,	82.2	lbs.
Temperature of feed-water,	205.3	deg.
Temperature of escaping gases,	445	deg.
Percentage of moisture in steam,	0.2	per cent.
Water per pound of coal,	8.39	lbs.
Water per pound of coal from and at 212 degrees,	8.76	lbs.
Water per pound of combustible from and at 212 degrees,	9.48	lbs.

The test on Boiler No. 39 shows the performance of a small horizontal tubular boiler using Walston bituminous coal. This boiler is not one from which to expect high results. Being deficient in heating surface, a considerable loss occurs from the heat of the waste gases, although the rate of combustion is comparatively low. The result of this test is similar, though less favorable, to that obtained on Boiler No. 34, which used the same kind of coal.

Boiler No. 40.

Kind of boiler,	Horizontal return tubular.
Number used,	One.
Horse-power (basis 12 square feet),	One hundred and five.
Kind of coal,	George's Creek Cumberland.
Age,	Several years.

Boiler No. 40 is of the ordinary horizontal tubular type, the general features of which, and the manner in which it is set, are shown in the following cut. The setting is so arranged that air is supplied above the fuel in two currents, one above and one below the burning gases. The first supply comes from the bridge wall, which is hollow, and it is discharged through perforations in the rear face of the wall near the top. The other current is supplied first to a second hollow wall placed a short distance behind the bridge wall, and suspended, so to speak, from the shell of the boiler, thereby causing the products of combustion to pass beneath it. The air emerges through perforations in the front face of this wall near the bottom.

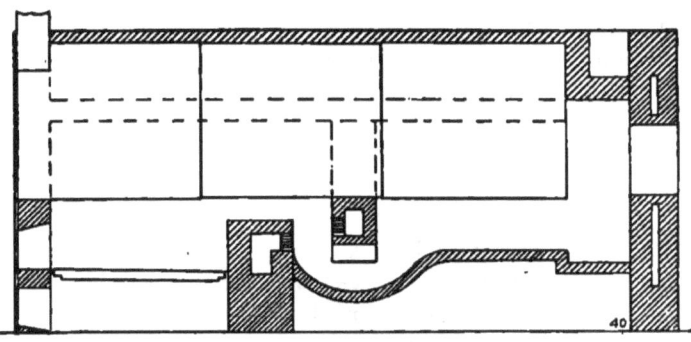

BOILER NO 40, LONGITUDINAL SECTION

Dimensions of Boiler No. 40.

Diameter of shell,	60	in.
Length between heads and length of tubes,	17	ft.
Number of tubes, 3½ inches outside diameter,	76	
Area of heating surface,	1,262	sq. ft.
Area of grate surface,	23.7	sq. ft.
Area through tubes,	4.4	sq. ft.
Width of air spaces and metal bars in grates,	Air 1-2 in., metal 3-8 in.	
Distance of grate to shell,	24	in.
Distance of flat bridge to shell,	9	in.
Ratio of heating surface to grate surface,	53.1 to 1	
Ratio of grate surface to tube area,	5.4 to 1	

Results of Test, Boiler No. 40.

Test No. 81.

Manner of start and stop,	Ordinary, with preliminary heating.	
Kind of run,	Continuous.	
Duration,	9.7	hrs.
Coal per hour,	3,140	lbs.
Percentage of ash,	7.5	per cent.
Water evaporated,	34,116	lbs.
Coal per hour,	323	lbs.
Coal per hour per square foot of grate,	13.6	lbs.
Water per hour,	3,499.4	lbs.
Water per hour per square foot of heating surface,	2.8	lbs.
Horse-power developed,	108.1	H. P.
Boiler pressure,	61.7	lbs.
Temperature of feed-water,	178.7	deg.
Temperature of escaping gases,	413	deg.
Draught suction,	0.15	in.
Water per pound of coal,	10.83	lbs.
Water per pound of coal from and at 212 degrees,	11.54	lbs.
Water per pound of combustible from and at 212 degrees,	12.47	lbs.

The test on Boiler No. 40 shows the performance of a horizontal tubular boiler using Cumberland bituminous coal. The conditions under which this test was made are all favorable to the production of a high result, and the desired end was realized in a notable degree, the evaporation being 12.47 pounds from and at 212 degrees per pound of combustible. The ratio of heating surface to grate surface is ample, being 53.1 to 1, the coal was burned at a sufficient rate for good combustion, a supply of air was introduced above the fuel in such a way as to thoroughly mix with the burning gases, and the heat wasted at the chimney was noticeably small for bituminous coal. The character of the combustion, as viewed through a peek-hole at the back end of the boiler, was excellent, and the appearance of the brick work at the rear part of the setting indicated a high degree of heat. There was a marked absence of dense black smoke discharged from the chimney.

Boiler No. 41.

Kind of boiler,	"Double-deck," hor. ret. tub.
Number used,	One.
Horse-power (basis 15 square feet),	Eighty-five.
Kind of coal,	Bituminous Cumberland.
Age,	One month.

Boiler No. 41 is a so-called "double-deck" boiler, arranged and set in the manner shown in the following cuts. It differs from the ordinary form of horizontal return tubular boiler in having two connecting shells, one above the other. The lower shell is completely filled with tubes. The upper shell is provided simply to furnish steam

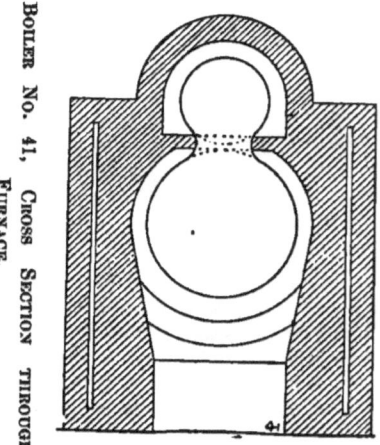

Boiler No. 41, Cross Section through Furnace.

room, and the water line is carried up to about the middle point in this shell. The area of heating surface for a given amount of grate surface, and the area through the tubes, is much larger in this form of boiler than in the ordinary type. In this respect the boiler, from an engineering point of view, possesses its chief characteristic. The brick setting is arched over the top of the drum, as shown in the cut, and the whole of the shell is exposed to the heat of the escaping gases on their way to the chimney. The boiler is thus provided with a small amount of steam-heating surface.

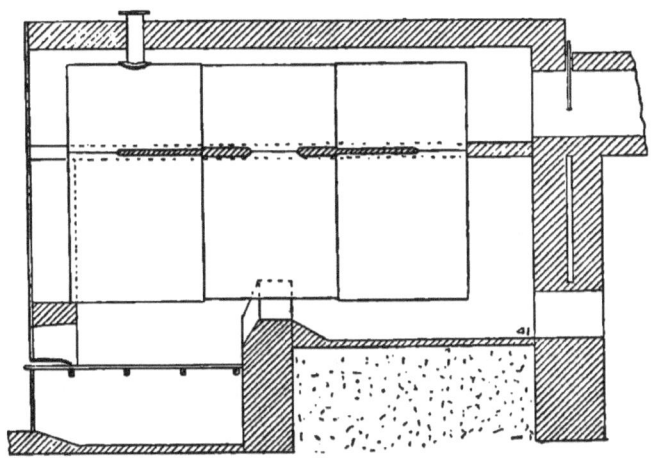

BOILER NO. 41, LONGITUDINAL SECTION.

Dimensions of Boiler No. 41.

Diameter of lower shell,	54 in.
Diameter of upper shell,	32 in.
Length of shells and tubes,	12 ft.
Number of tubes (3 inches diameter),	118
Area of water-heating surface,	1,221 sq. ft.
Area of steam-heating surface,	60 sq. ft.
Area of grate surface,	20 sq. ft.
Area through tubes,	4.9 sq. ft.
Ratio of water-heating surface to grate,	61 to 1
Ratio of steam-heating surface to grate,	3 to 1
Ratio of grate to tube area,	4.1 to 1

Results of Tests, Boiler No. 41.

	Test No. 82.
Manner of start and stop,	Running.
Kind of run,	Continuous.
Duration,	12 hrs.
Coal consumed,	1,912 lbs.
Percentage of ash,	7 per cent.
Water evaporated,	19,991 lbs.
Coal per hour,	159.3 lbs.
Coal per hour per square foot of grate,	7.96 lbs.
Water per hour,	1,665.9 lbs.
Water per hour per square foot of water-heating surface,	1.36 lbs.
Horse-power developed,	50.4 H. P.
Boiler-pressure,	81 lbs.
Temperature of feed-water,	197 deg.
Temperature of escaping gases,	322 deg.
Percentage of moisture in steam,	0.5 per cent.
Water per pound of coal,	10.45 lbs.
Water per pound of coal from and at 212 degrees,	10.98 lbs.
Water per pound of combustible from and at 212 degrees,	11.81 lbs.

The results of the test on Boiler No. 41 are of interest in showing the effect produced by the large extent of heating surface which the double-deck type of boiler contains. The evaporation per pound of combustible from and at 212 degrees, which is 11.81, is higher than that obtained from some horizontal return tubular boilers of the ordinary kind; but at the same time, it is inferior to the best results from those boilers. The rate of combustion, which is 7.96 pounds per square foot of grate per hour, is probably too low for the best results on a boiler having such a large amount of heating surface, and the conditions of the tests are not in this respect the most favorable. Taking into account simply the evaporative result, it would appear that the increased surface over a well arranged return tubular boiler of the ordinary type, produced no material benefit. That the large area of heating surface was efficient in absorbing the heat is seen in the low temperature of the escaping gases, which was 322 degrees, this being below the temperature of the water in the boiler. The steam which the boiler generated contained 0.5 of one per cent. of moisture as indicated by a barrel calorimeter. From this it appears that the steam-heating surface in the upper shell was not sufficient to superheat the steam, although it may have acted favorably in reducing the amount of moisture present to the low figure named.

Boiler No. 42.

Kind of boiler, ... "Double-deck" hor. ret. tub.
Number used, ... Three.
Horse-power (collective, basis 15 square feet), Three hundred and ninety.
Age, ... New.

Boiler No. 42 embraces a plant of three 60-inch "double-deck" boilers set in one battery of brick-work, as indicated in cross section in the following cut. The longitudinal section is of the same general form as that shown in the cut of Boiler No. 41. This boiler, like the preceding one, has a much larger area of heating surface and tube area for a given size of grate than the ordinary tubular boiler, and this is its characteristic feature.

BOILER No. 42, CROSS SECTION THROUGH FURNACES.

Dimensions of Boiler No. 42.

Diameter of lower shell, ... 60 in.
Diameter of upper shell, ... 34 in.
Length of shell and tubes, ... 15 ft.
Number of tubes (collective) 3 inches outside diameter, . 420
Area of water-heating surface, ... 5,850 sq. ft.
Area of steam-heating surface, ... 300 sq. ft.

BOILER No. 42.

Area of grate surface,	90 sq. ft.
Area through tubes,	17.3 sq. ft.
Area through chimney,	11.1 sq. ft.
Height of chimney,	100 ft.
Ratio of water-heating surface to grate surface,	65 to 1
Ratio of steam-heating surface to grate surface,	3.3 to 1
Ratio of grate to tube area,	5.2 to 1
Ratio of grate to chimney area,	8.1 to 1

Results of Tests, Boiler No. 42, (Average of three.)

	Test No. 83.	Test No. 84.
Kind of coal,	Anthracite Lackawanna, Broken.	Bituminous, Geo.'s Creek Cumberland.
Manner of start and stop and kind of run,	Ordinary.	Ordinary.
Duration, hrs.	12	11.8
Coal consumed, dry (including wood equivalent), lbs.	15,385	11,646
Percentage of ash, per cent.	12.3	6.7
Water evaporated, lbs.	141,574	129,348
Coal per hour, lbs.	1,282.1	987
Coal per hour per square foot of grate, lbs.	14.25	10.97
Water per hour, lbs.	11,797.8	10,961.7
Water per hour per square foot of water-heating surface, lbs.	2.02	1.87
Horse-power developed, H. P.	359	331.8
Boiler pressure, lbs.	82.5	81.7
Temperature of feed-water, deg.	199.3	205
Temperature of escaping gases, deg.	392	389
Percentage of moisture in steam, per cent.	—	0.5
Water per pound of coal, lbs.	9.20	11.08
Water per pound of coal from and at 212 degrees, lbs.	9.65	11.55
Water per pound of combustible from and at 212 degrees, lbs.	11.11	12.42

The tests on Boiler No. 42 were made in the same series as those of Boiler No. 3, one set of boilers being used when the other set was idle, and both being employed for the same work. A comparison of Tests No. 5 and No. 83 may be made to show the efficiency of the two types of boilers. Both tests were made with coal from the same cargo. The double-deck boiler secured a somewhat better result than the common boiler, the figures per pound of combustible being 11.11 and 10.73. The difference is 3.5 per cent. The reason for this result is seen in the reduced quantity of waste heat. The temperature

of the escaping gases in the two cases is 482 degrees and 392 degrees, respectively. The reduction is evidently due in a measure to the increased amount of heating surface which the double-deck boiler provides. There is 5,850 square feet of surface in this boiler, against only 4,056 square feet in the common boilers. While an improvement might be expected from this cause, the whole of the improvement cannot be attributed to the difference in the type of boiler, because the double-deck boilers were new and clean, while the common boilers were old and presumably somewhat inefficient from long usage. It is quite probable that the whole of this small improvement (which, as noted, is only 3.5 per cent.) may be attributed to the difference in age and condition of the heating surfaces, rather than to the change in the type of boiler.

The results of Tests No. 83 and No. 84 may be compared to show the relative economy of Lackawanna coal, broken size, and George's Creek Cumberland coal. The bituminous coal, considered by itself, gave exceptionally good results. The large draught area through the tubes appears to be a favorable condition in cases like this, where the gaseous element of combustion forms so prominent a feature. If this be so, the comparison between the economy of the two fuels shows a larger difference in favor of the bituminous coal than would be shown on some different type of boiler. Taking the figures as they stand, however, there is a difference of $11.55 - 9.65 = 1.90$ pounds of water in favor of the performance of the bituminous coal, which is a gain of nearly 20 per cent. over that of the anthracite coal, and this gain represents a saving in the quantity of bituminous coal required to do a given amount of work of 16.4 per cent.

In comparing the two kinds of fuel used on Tests No. 83 and No. 84, a marked difference is to be observed in the quantity of ash. The percentage of ash in the anthracite coal is 12.3 per cent. and that in the bituminous coal 6.7 per cent.

Boiler No. 43.

Kind of boiler,	"Double-deck" hor. ret. tub.
Number used,	One.
Horse-power (basis 15 square feet),	One hundred.
Age,	New.

Boiler No. 43 is of the double-deck type having the general features shown in the following cuts. It has two shells, one completely filled with tubes and the other serving principally as a steam drum. The brick setting is arranged so as to admit air through the perforations in the side walls at the back end of the furnace and perforations in the top of the bridge wall. The air supplied in this manner is first passed through ducts extending back and forth through the walls, whereby it is to some extent heated.

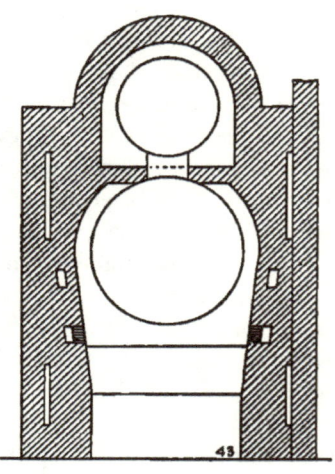

BOILER No. 43, CROSS SECTION THROUGH FURNACE.

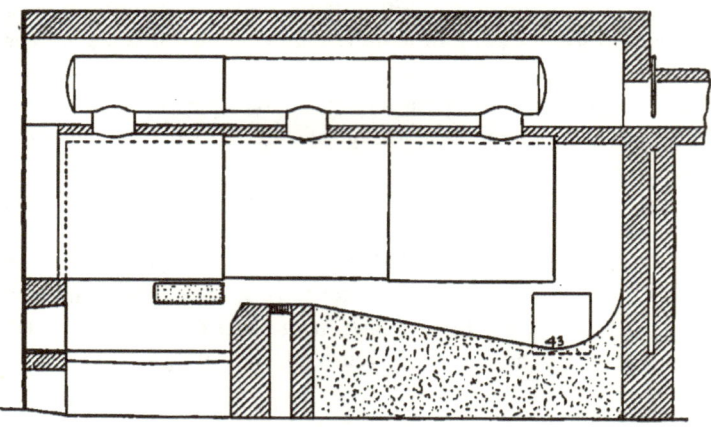

BOILER No. 43, LONGITUDINAL SECTION.

Dimensions of Boiler No. 43.

Diameter of lower shell,	54 in.
Diameter of upper shell,	36 in.
Length of shell and tubes,	15 ft.
Number of tubes 3 inches outside diameter,	114
Area of water-heating surface,	1,521 sq. ft.
Area of steam-heating surface,	88 sq. ft.
Area of grate surface,	22.5 sq. ft.
Area through tubes,	9.4 sq. ft.
Area through chimney,	7.1 sq. ft.
Height of chimney,	70 ft.
Width of air spaces and metal bars in grates,	3- 8 in.
Ratio of water-heating surface to grate surface,	67.6 to 1
Ratio of steam-heating surface to grate surface,	3.9 to 1
Ratio of grate surface to tube area,	4.8 to 1

Results of Tests, Boiler No. 43.

	Test No. 85.	Test No. 86.	Test No. 87.	Test No. 88.
Kind of fuel,	Mixture 1 pt. Nova Scotia Culm, 3 pts. Pea and Dust.	Delaware and Lackawanna Broken Anthracite.	George's Creek Cumberland Bituminous.	Nova Scotia Culm.
Manner of start and stop and kind of run,	Av. 3 days. Ordinary.	Av. 2 days. Ordinary.	Av. 2 days. Ordinary.	One day. Ordinary.
Duration, hrs.	11.5	11.5	11.5	9.5
Coal consumed, dry (including wood equivalent,) lbs.	2,983	2,485	2,131	2,390
Percentage of ash, per cent.	14.7	15.5	6.7	10.4
Water evaporated, lbs.	24,671	21,091	21,137	18,938
Coal per hour, lbs.	259.3	216.1	185.4	247.7
Coal per hour per square foot of grate, lbs.	11.5	9.4	8.2	11.2
Water per hour, lbs.	2,145	1,834	1,838	1,994
Water per hour per square foot of water-heating surface, lbs.	1.4	1.2	1.2	1.3
Horse-power developed, H. P.	64.3	55	55	59.8
Boiler pressure, lbs.	75.3	74	74.2	71.7
Temperature of feed-water, deg.	206	211	211	210
Temperature of escaping gases, deg.	339	335	348	348
Moisture in steam by calorimeter, per cent.	1.2	–	–	–
Water per pound of coal, lbs.	8.27	8.48	9.92	7.93
Water per pound of coal from and at 212 degrees, lbs.	8.59	8.78	10.26	8.21
Water per pound of combustible from and at 212 degrees, lbs.	10.07	10.39	10.99	9.19

NOTE.—The coals when fired contained the following percentages of moisture: Mixed fuel, 6 per cent.; Cumberland, 2.5 per cent.; Nova Scotia Culm, 7 per cent.

The tests on Boiler No. 43 embrace a series made to determine the relative economy of different kinds of coal. The best evaporative result was obtained with George's Creek Cumberland coal, this being 10.26 pounds of water from and at 212 degrees per pound of coal. Compared with the other results this is 17 per cent. better than that obtained with anthracite coal, 19 per cent. better than that given by the mixed coal, and 25 per cent. better than that given by the Nova Scotia culm. Based on the cost of fuel required to produce one day's supply of steam, say 30,000 pounds from and at 212 degrees, using the prices of coal which ruled at the time of the tests, the various results are as follows:

	Mixture, Pea and Dust 3 parts, Nova Scotia Culm 1 part.	Delaware and Lackawanna Anthracite Broken.	George's Creek Cumberland Bituminous.	Nova Scotia Culm.
Cost of 2,240 pounds coal,	$4 02	$5 90	$6 75	$3 80
Cost of coal for 30,000 pounds steam,	6 31	9 18	8 91	7 14

From these figures it is seen that the best result in point of cost of fuel was obtained with the mixed fuel, and this is represented by an expenditure of $6.31 per day. Comparing the best grades of coal which were tested, viz: the Anthracite and Cumberland, it appears that the highest priced coal per ton, that is, the Cumberland, owing to its better evaporative performance, was in reality the cheapest, the cost per day's run being $0.27 less than that of anthracite coal.

During Test No. 86 the air passages in the walls were closed and no air entered the furnace above the fuel except that which found its way through cracks in the setting, and through the registers in the fire door.

During one day's run on Test No. 87, which was made with Cumberland coal, the coal was wet after being weighed, the amount of water used being 5 per cent. of the weight of the coal. On the day when the coal was wet, the quantity of water evaporated by the boiler with the same weight of dry coal, was increased 3 per cent.

The individual results of the tests made with the standard grades of coal, viz: Tests No. 86 and No. 87, are not so high as the best obtained from other boilers of similar type. Boiler No. 41 with Cumberland coal gave an evaporation per pound of combustible from and at 212 degrees of 11.81 pounds, and Boiler No. 42 with similar coal, gave 12.42 pounds. The last named boiler with anthracite coal gave 11.11 pounds. All of these figures are higher than the corresponding results obtained on Boiler No. 43. From the results of a subsequent test on Boiler No. 43, which was made when it was working to a greater capacity, it appears that the economy of the boiler was effected by the low rate of combustion under which it was worked. On the test referred to, which was made with a mixture of pea and dust coal and Nova Scotia culm, having the same proportions as those used on Test No. 85, the rate of combustion was increased to 15.5 pounds of coal per square foot of grate per hour. The evaporation from and at 212 degrees per pound of coal was 9.18 pounds, and per pound of combustible 10.77 pounds, and these quantities are about 7 per cent. higher than those obtained on the earlier tests. The temperature of the escaping gases on the last test was 375 degrees.

Boiler No. 44.

Kind of boiler, ... "Double deck," hor. ret. tub.
Number used, ... Four.
Horse-power (collective, basis 15 square feet), Three hundred and twenty.
Kind of coal, ... Eureka, Cumberland, Bituminous.
Age, ... New.

Boiler No. 44 consists of a plant of four double-deck boilers set in one battery of brick work in the general manner shown in the cuts of Boilers No. 41 and No. 42. Like those referred to, this boiler has two shells, one above the other. The lower one is completely filled with tubes, and by this means the area of heating surface and the area through the tubes is much larger in proportion to the grate surface than is found in ordinary tubular boilers.

BOILER No. 44.

Dimensions of Boiler No. 44.

Diameter of lower shell,	54 in.
Diameter of upper shell,	32 in.
Length of shells and tubes,	12 ft.
Number of tubes (collective) 3 inches outside diameter,	472
Area of water-heating surface,	4,972 sq. ft.
Area of steam-heating surface,	320 sq. ft.
Area through tubes,	19.5 sq. ft.
Area of grate surface,	80 sq. ft.
Width of air spaces and metal bars in grates,	3-8 in.
Ratio of water-heating surface to grate surface,	62.1 to 1
Ratio of steam-heating surface to grate surface,	4 to 1
Ratio of grate to tube area,	4.1 to 1

Results of Tests, Boiler No. 44. (Average of two.)

Test No. 80.

Manner of start and stop and kind of run,		Ordinary.
Duration,	12.25	hrs.
Coal consumed, dry (including wood equivalent),	9,000	lbs.
Percentage of ash,	9.7	per cent.
Water evaporated,	92,355	lbs.
Coal per hour,	741.8	lbs.
Coal per hour per square foot of grate,	9.3	lbs.
Water per hour,	7,539.2	lbs.
Water per hour per square foot of water heating surface,	1.5	lbs.
Horse-power developed,	231.4	H. P.
Boiler pressure,	82	lbs.
Temperature of feed-water,	190	deg.
Temperature of escaping gases,	375	deg.
Water per pound of coal,	10.26	lbs.
Water per pound of coal from and at 212 degrees,	10.86	lbs.
Water per pound of combustible from and at 212 degrees,	12.03	lbs.

The economical result obtained on Boiler No. 44 exceeds 12 pounds of water from and at 212 degrees per pound of combustible. This is a creditable performance, though it is no higher than that given by many boilers of the ordinary horizontal tubular type, a fact which shows that some of the heating surface in this form of boiler fails to be utilized.

Boiler No. 45.

Kind of boiler,	"Double deck" hor. ret. tub.
Number used,	One.
Horse-power (basis 15 square feet),	One hundred and fifty.
Kind of coal,	Delaware and Lackawanna anthracite, broken.
Age,	New.

Boiler No. 45 is a double deck boiler similar in all features except size to Boilers No. 41 and No. 42, and the manner in which it is arranged and set is shown in the cuts of these boilers.

Dimensions of Boiler No. 45.

Diameter of lower shell,	66 in.
Diameter of upper shell,	36 in.
Length of shells and tubes,	15 ft.
Number of tubes 3 inches outside diameter,	172
Area of water-heating surface,	2,235 sq. ft.
Area of steam-heating surface,	70 sq. ft.
Area through tubes,	7.1 sq. ft.
Area of grate surface,	36.6 sq. ft.
Width of air spaces and metal bars in grates,	Air 5-8 in., metal 1-2 in.
Ratio of water-heating surface to grate surface,	60.9 to 1
Ratio of steam-heating surface to grate surface,	2 to 1
Ratio of grate surface to tube area,	5.2 to 1

Results of Tests, Boiler No. 45. (Average of two.)

	Test No. 90.
Manner of start and stop and kind of run,	Ordinary.
Duration,	11.7 hrs.
Coal consumed (including wood equivalent),	6,237 lbs.
Percentage of ash,	8.1 per cent.
Water evaporated,	56,388 lbs.
Coal per hour,	530.8 lbs.
Coal per hour per square foot of grate,	14.5 lbs.
Water per hour,	4,799 lbs.
Water per hour per square foot of water-heating surface,	2.1
Horse-power developed,	155.1 H. P.
Boiler pressure,	79 lbs.
Temperature of feed water,	132 deg
Temperature of escaping gases,	373 deg.
Percentage of moisture in steam,	0.3 per cent.
Water per pound of coal,	9.04 lbs.
Water per pound of coal from and at 212 degrees,	9.76 lbs.
Water per pound of combustible from and at 212 degrees,	11.00 lbs.

The tests on Boiler No. 45 were made with damper wide open, and therefore under conditions of maximum capacity. The chimney was of ample size, and its height was 70 feet. The boiler power developed amounted to 155 horse power, or a trifle more than the nominal capacity rated on 15 square feet of heating surface per horse power. Although the boiler worked up to the capacity noted, it could not properly be called a 150 horse power boiler, unless provided with a chimney of such height that it would produce that power with considerably less than the full draught. The low temperature of the escaping gases shows that the large area of heating surface served to absorb nearly the whole of the available heat of the products of combustion, and the evaporation per pound of combustible from and at 212 degrees, which was 11 pounds, is a favorable result.

Boiler No. 46.

Kind of boiler, ... Hor. ret. tub. (double-deck).
Number used, ... Four.
Horse power (collective, basis 15 square feet), Two hundred and ninety.
Kind of coal, ... Cumberland.
Age, ... Six years.

Boiler No. 46 embraces a plant of four double deck horizontal tubular boilers set in one battery of brick work. The general features of the boiler and its setting are shown in the cuts of Boilers No. 41 and No. 42. It is provided with a flue heater, consisting of vertical cast-iron pipes, similar to that used with Boiler No. 33, the location of which, with reference to boilers and chimney, is shown in ground plan in the following cut. This heater has about one-half as much surface as the total heating surface of the boilers.

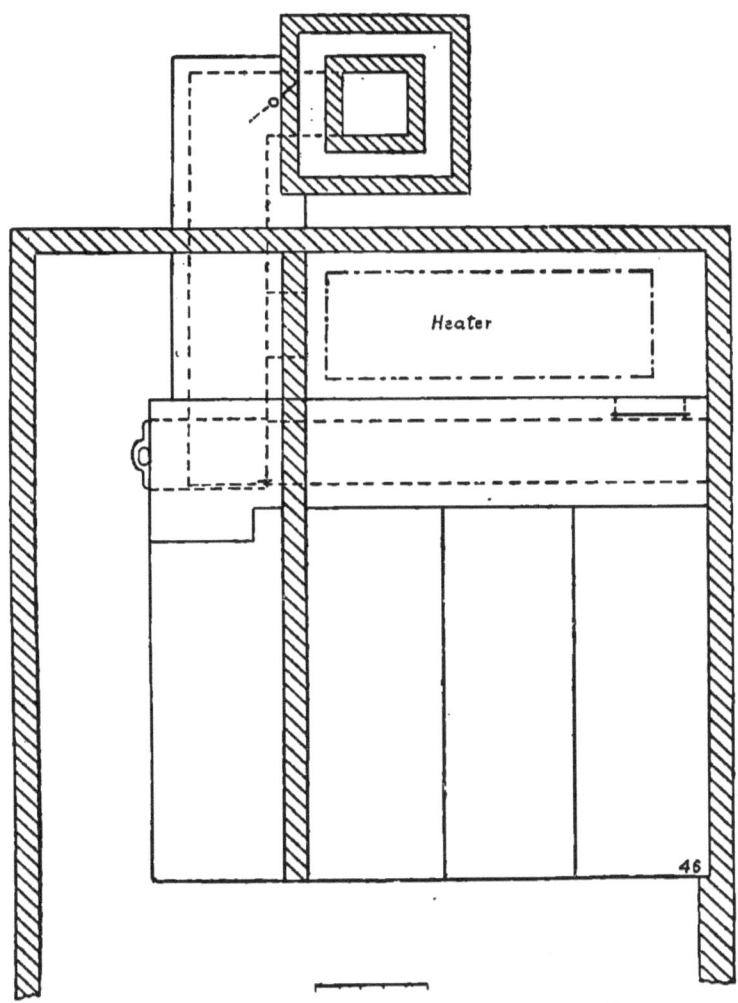

Boiler No. 46, Ground Plan showing Location of Boilers and Flue Heater.

BOILER No. 46.

Dimensions of Boiler No. 46.

Diameter of lower shell,	48 in.
Diameter of upper shell,	30 in.
Length of shells and tubes,	14 ft.
Number of tubes (collective) 3 inches outside diameter,	320
Area of water-heating surface,	4,058 sq. ft.
Area of steam-heating surface,	280 sq. ft.
Area of grate surface,	70 sq. ft.
Area through tubes,	13.2 sq. ft.
Area through flue,	9 sq. ft.
Height of chimney,	80 ft.
Width of air spaces and metal bars in grates,	Air 5-8 in., metal 1 in.
Ratio of water-heating surface to grate surface,	58 to 1
Ratio of steam-heating surface to grate surface,	4 to 1
Ratio of grate surface to tube area,	5.3 to 1
Ratio of grate surface to flue area,	7.8 to 1
Area of heating surface in flue heater,	1,920 sq. ft.

Results of Tests, Boiler No. 46.

	Test No. 91.	Test No. 92.
	Heater in use.	Heater not in use.
Manner of start and stop and kind of run,	Ordinary.	Ordinary.
Duration, hrs.	12.5	12.5
Coal consumed, dry (including wood equivalent), lbs.	10,659	10,946
Percentage of ash, per cent.	7.6	7
Water evaporated, lbs.	106,598	102,261
Coal per hour, lbs.	852.7	875.6
Coal per hour per square foot of grate, lbs.	12.2	12.5
Water per hour, lbs.	8,527.8	8,180.8
Water per hour per square foot of water-heating surface, lbs.	2.1	2
Horse-power developed, H. P.	289.9	277.8
Boiler pressure, lbs.	77	75
Temperature of feed-water entering heater, deg.	79	–
Temperature of feed-water entering boiler, deg.	145	79
Temperature of escaping gases leaving boiler, deg.	361	342
Temperature of escaping gases leaving heater, deg.	254	–
Draught suction, in.	0.17	0.22
Water per pound of coal, lbs.	10.00	9.34
Water per pound of coal from and at 212 degrees, lbs.	–	10.93
Water per pound of combustible from and at 212 degrees, lbs.	–	11.78

NOTE.— The coal when fired contained 5 per cent. of moisture.

The test on Boiler No. 46 had for a principal object the determination of the economy of using a flue heater with horizontal tubular boilers, which were already provided with a large area of heating surface, the ratio of which to grate surface was 58 to 1. The heater reduced the temperature of the gases 361° — 254° = 107 degrees, and increased the temperature of the feed-water 145° — 79° = 66 degrees, and secured thereby an increase in the evaporation per pound of coal amounting to 7 per cent. If the results of Test No. 91 are applied to the plant as a whole, the evaporation per pound of combustible from and at 212 degrees becomes 12.68 pounds. Looked at in this way, the performance is exceedingly high for this or any other type of boiler, and it shows what may be attained by employing an ample area of heating surface, arranged in such a manner as to properly absorb the heat. Without the heater, the economy obtained is below the best performance of ordinary horizontal tubular boilers.

Boiler No. 47.

Kind of boiler,	Plain cylinder.
Number used,	Four.
Horse-power (collective, basis 6 sq. ft.),	Two hundred and twenty.
Kind of coal,	Anthracite, chestnut No. 2.
Age,	Ten years.

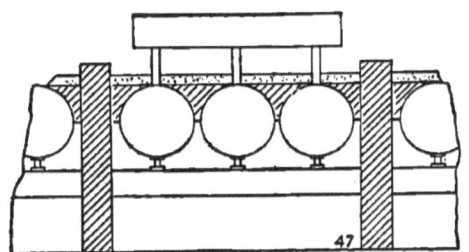

BOILER NO. 47, CROSS SECTION THROUGH FURNACE.

Boiler No. 47 embraces four sets of cylinder boilers arranged in the manner shown in the following cuts. Each boiler consists of three shells placed over a single furnace, the lower portions of which are exposed their entire length to the heat of the products of combustion. The upper portions are covered with brick work. These boilers have a ratio of only 7.5 square feet of heating surface to one of grate. In this respect they possess their special characteristic. The ratio in a horizontal tubular boiler, is seldom below 25 to 1, and the proportion of water-heating surface in the vertical boiler is seldom below 15 to 1.

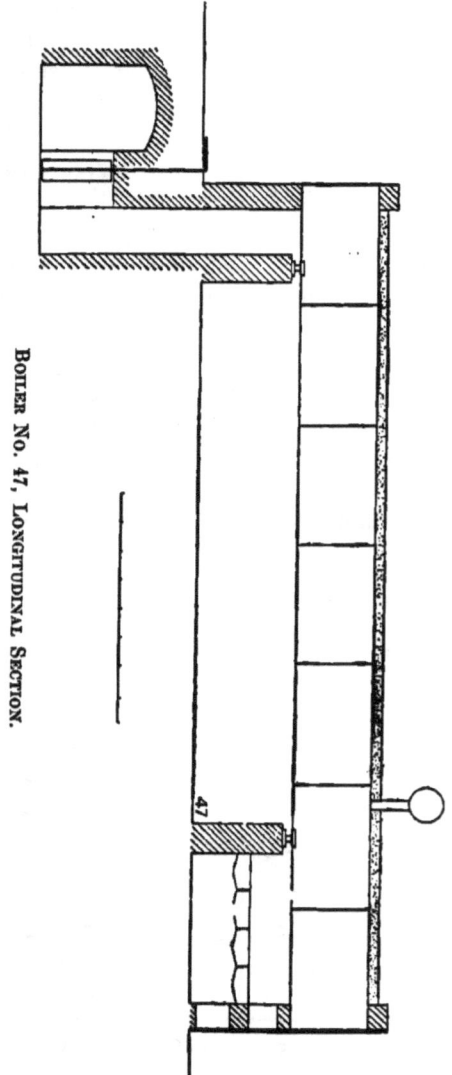

BOILER NO. 47, LONGITUDINAL SECTION.

Dimensions of Boiler No. 47.

Diameter of shells (3 in each),	30	in.
Length of shells,	30	ft.
Area of heating surface,	1,320	sq. ft.
Area of grate surface,	17.5	sq. ft.
Area through flue,	10.5	sq. ft.
Height of chimney,	120	ft.
Width of air spaces and metal bars in grates,	3-8	in.
Distance of grate to shell,	16	in.
Ratio of heating surface to grate surface,	7.5	to 1
Ratio of grate surface to flue area,	16.7	to 1

Results of Tests, Boiler No. 47.

	Test No. 93.	Test No. 94.
Manner of start and stop and kind of run,	Ordinary.	Ordinary.
Duration, ... hrs.	10.7	11.5
Coal consumed, dry (including wood equivalent), ... lbs.	14,061	12,264
Percentage of ash, ... per cent.	14	14.7
Water evaporated, ... lbs.	98,256	83,416
Coal per hour, ... lbs.	1,308	1,066.4
Coal per hour per square foot of grate, lbs.	7.4	6.1
Water per hour, ... lbs.	9,140.1	7,253.6
Water per hour per square foot of heating surface, ... lbs.	6.9	5.5
Horse-power developed, . . . H. P.	272	240
Boiler pressure, ... lbs.	72.7	71.2
Temperature of feed-water, . . deg.	207	93
Temperature of escaping gases, . deg.	650	567
Water per pound of coal, . . . lbs.	6.99	6.80
Water per pound of coal from and at 212 degrees, ... lbs.	7.27	7.89
Water per pound of combustible from and at 212 degrees, ... lbs.	8.44	9.22

NOTE.— The coal when fired contained 5 per cent. of moisture. On Test No. 94 the ash-pit doors were partly closed.

The tests on Boiler No. 47 exhibit the performance of plain cylinder boilers using one of the small grades of anthracite coal and working under two different rates of combustion. One noticeable feature in the results is the high rate of evaporation which occurred, being 6.9 pounds per square foot of heating surface per hour in Test No. 93, and 5.5 pounds in Test No. 94. Another is the high temperature of the escaping gases, which was 650 degrees in the first test, and 567 degrees in the second test. Both of these are due to the relatively small amount of the heating surface. In view of the low results which are often obtained on more modern boilers, the performance here shown, must, under the circumstances, be considered remarkable. Looking at Test No. 94, the rate of evaporation exceeds anything ordinarily obtained from boilers of approved type, and the temperature of the escaping gases is correspondingly high, but the evaporative result, viz., 9.22 pounds of water from and at 212 degrees per pound of combustible, is not more than 20 per cent. below the highest that can ordinarily be obtained from the horizontal tubular boiler with the class of fuel used.

Boiler No. 48.

Kind of boiler,	Plain cylinder.
Number used,	One.
Horse-power (basis 6 sq. ft.),	Seventy-seven.
Age,	Several years.

Boiler No. 48 is of the plain cylinder type, arranged and set in the general manner shown in the cut of Boiler No. 47. Here four shells instead of three are used, and the setting of the boiler is independent of other boilers. The proportion of heating surface to grate surface is somewhat larger than in Boiler No. 47, being 10.9 to 1, though very far below that of good practice in approved boiler work.

Dimensions of Boiler No. 48.

Number of shells,	4
Diameter of each shell,	30 in.
Length of each shell,	30.5 ft.
Area of heating surface,	464 sq. ft.
Area of grate surface,	42.7 sq. ft.
Area through flue,	6 sq. ft.
Height of chimney,	75 ft.
Width of air spaces and metal bars in grates,	Air 5-16 in., metal 3-8 in.
Ratio of heating surface to grate surface,	10.9 to 1
Ratio of grate surface to flue area,	7.1 to 1

Results of Tests, Boiler No. 48.

		Test No. 95.	Test No. 96.
Kind of coal,		Coke.	Anthracite Pea.
Manner of start and stop,		Thin fire.	Thin fire.
Kind of run,		Factory.	Factory.
Duration,	hrs.	10.5	11.5
Coal consumed, dry	lbs.	5,774	4,770
Percentage of ash,	per cent.	7.7	13.2
Water evaporated,	lbs.	30,140	27,258
Coal per hour,	lbs.	549.9	414.8
Coal per hour per square foot of grate,	lbs.	12.8	9.7
Water per hour,	lbs.	2,870.5	2,370.3
Water per hour per square foot of heating surface,	lbs.	6.2	5.1
Horse-power developed,	H. P.	99.7	82.4
Boiler pressure,	lbs.	82	79
Temperature of feed-water,	deg.	39	38
Temperature of escaping gases,	deg.	Above 600	Above 600
Water per pound of coal,	lbs.	5.22	5.71
Water per pound of coal from and at 212 degrees,	lbs.	6.37	6.96
Water per pound of combustible from and at 212 degrees,	lbs.	6.87	7.97

The tests on Boiler No. 48 were made to determine the relative economy of anthracite pea coal and gas house coke, and, incidentally, the general economy of the boiler. The evaporation of water per pound of coke is 8 per cent. less than the evaporation per pound of coal, and this relation is also that of the cost of fuel in the two cases, the price per pound being the same. The effect of the deficiency of heating surface in this boiler is seen in the exceedingly high temperature of the gases, and the consequent low degree of economy. Comparing the result of Test No. 96 made on this boiler with pea coal, with Test No. 36 made on horizontal tubular boiler No. 15 with similar fuel, it appears that the evaporation per pound of combustible is 20.1 per cent. less than that obtained in the case of the improved form of boiler. Comparing also Test No. 95 made on this boiler when using coke, with No. 67 made on tubular boiler No. 31 with the same kind of fuel, it appears that the cylinder boiler gave an evaporation of 32 per cent. less water per pound of combustible, than that obtained with the tubular boiler.

Boiler No. 49.

Kind of boiler,	Plain cylinder.
Number used,	One.
Horse-power (basis 10 square feet),	Thirty-nine.
Age,	Several years.

Boiler No. 49 is a plain cylinder boiler, the general features of which are shown in the cuts of Boiler No. 47. The peculiarity of this type of boiler is the small proportion of heating surface to grate surface, which here was 10.9 to 1.

Dimensions of Boiler No. 49.

Diameter of shells, three in number,	30 in.
Length of shells,	30 ft.
Area of heating surface,	394 sq. ft.
Area of grate surface,	36.1 sq. ft.
Area through flue,	5.4 sq. ft.
Height of chimney,	120 ft.
Width of air spaces and metal bars in grates,	1-2 in.
Distance of grate to shell,	17 in.
Distance of flat bridge to shell,	4 in.
Ratio of heating surface to grate surface,	10.9 to 1.
Ratio of grate surface to flue area,	6.6 to 1.

Results of Tests, Boiler No. 49.

	Test No. 97.	Test No. 98.	Test No. 99.
Kind of coal,	Bituminous Cumberland.	Bituminous Cumberland.	Anthracite Chestnut No. 2.
Condition of fire at start and stop,	Thin fire.	Banked fire.	Ordinary with preliminary heating.
Kind of run,	Ordinary.	Ordinary.	Continuous.
Duration, ... hrs.	9.5	12.2	5.2
Coal consumed, dry (including wood equivalent), lbs.	2,579	3,338	2,014
Percentage of ash, per cent.	10	10.3	19.3
Water evaporated, lbs.	17,270	22,119	9,643
Coal per hour, lbs.	271.4	272.5	383.6
Coal per hour per square foot of grate, lbs.	7.5	7.5	10.6
Water per hour, lbs.	1,817.9	1,805.6	1,836.6
Water per hour per square foot of heating surface, lbs.	4.6	4.6	4.7
Horse-power developed, H. P.	60.8	60.8	62
Boiler pressure lbs.	80	80	83
Temperature of feed-water, deg.	78	90	90
Temperature of escaping gases, deg.	Melts zinc	–	–
Draught suction, in.	0.18	0.17	0.16
Water per pound of coal, lbs.	6.70	6.63	4.94
Water per pound of coal from and at 212 degrees, lbs.	7.86	7.70	5.75
Water per pound of combustible from and at 212 degrees, lbs.	8.74	8.59	7.03

The tests on Boiler No. 49 had for an object the determination of the general economy of this form of boiler, the relative economy of Cumberland coal and anthracite chestnut No. 2 coal, and the loss occasioned by banking fires. The results all show a very low degree of economy. That obtained on Test No. 97 is more than 25 per cent. inferior to the best results obtained on boilers of approved type. Attention needs only to be directed to the deficient heating surface, and the consequent large amount of waste heat in the gases, which in this case were hot enough to melt zinc, to account for the result. Comparing Test No. 99 with Test No. 97, the evaporation per pound of chestnut No. 2 coal is 26 per cent. less than that of Cumberland coal. The loss produced by banking a Cumberland coal fire, as indicated by Tests No. 97 and No. 98, appears to be 2 per cent.

Boiler No. 50.

Kind of boiler,	Galloway.
Number used,	Six.
Horse-power (collective, makers' rating),	One thousand, four hundred and forty.
Kind of coal,	George's Creek Cumberland.
Age,	Three months.

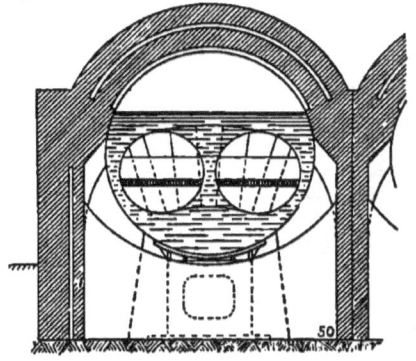

BOILER NO. 50, CROSS SECTION THROUGH FURNACES.

Boiler No. 50 consists of a plant of six Galloway boilers, having the general features shown in the following cuts. This is an internally fired boiler having two furnaces in each shell. The heating surface consists largely of vertical water tubes of conical shape, which extend from top to bottom of the large flue which passes from the combustion chamber at the end of the furnaces to the end of the boiler. The products of combustion, on leaving the furnaces, pass forward through this flue, thence backward beneath the shell, finally entering the chimney flue from the front end. In this boiler the ratio of heating surface to grate surface is much smaller than is generally found in horizontal tubular boilers of approved type.

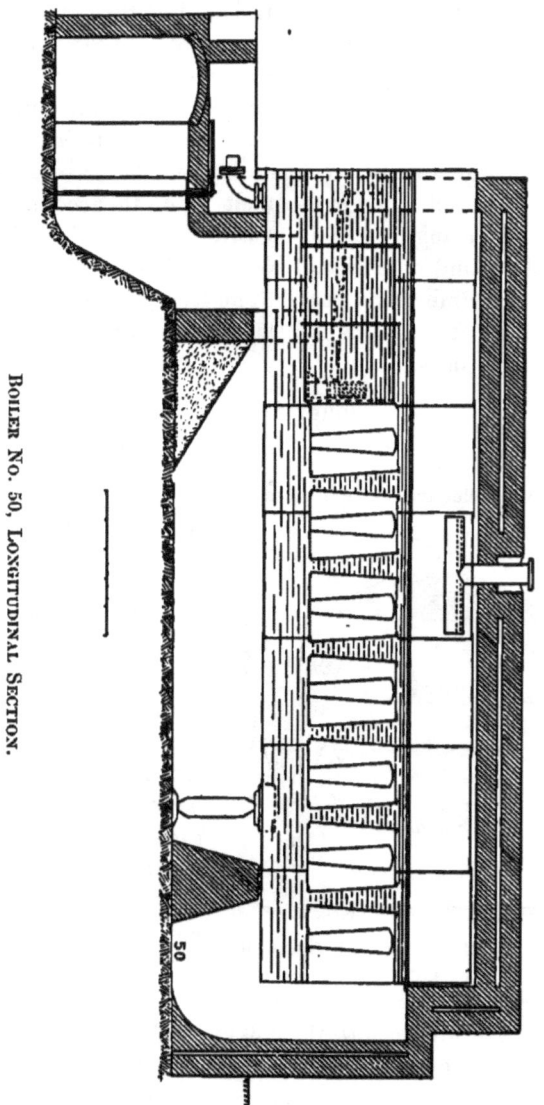

Boiler No. 50, Longitudinal Section.

Dimensions of Boiler No. 50.

Diameter of each shell,	84 in.
Length of shell,	28 ft.
Area of heating surface,	5,628 sq. ft.
Area of grate surface,	217.6 sq. ft.
Area through flue,	60 sq. ft.
Width of air spaces and metal bars in grates,	Air 1 in., metal 3-8 in.
Height of chimney,	230 ft.
Ratio of heating surface to grate surface,	25.9 to 1
Ratio of grate surface to flue area,	3.6 to 1

Results of Test, Boiler No. 50.

	Test No. 100.
Manner of start and stop,	Running with thin fire.
Kind of run,	Continuous.
Duration,	12.7 hrs.
Coal consumed, dry,	47,638 lbs.
Percentage of ash,	6.0 per cent.
Water evaporated,	370,448 lbs.
Coal per hour,	3,756.9 lbs.
Coal per hour per square foot of grate,	17.3 lbs.
Water per hour,	29,210.3 lbs.
Water per hour per square foot of heating surface,	5.2 lbs.
Horse-power developed,	1,024.3 H. P.
Boiler pressure,	51.2 lbs.
Temperature of feed-water,	36.7 deg.
Temperature of escaping gases,	533 deg.
Draught suction,	0.64 in.
Percentage of moisture in steam,	7 per cent.
Water per pound of coal,	7.78 lbs.
Water per pound of coal from and at 212 degrees,	9.40 lbs.
Water per pound of combustible from and at 212 degrees,	10.00 lbs.

The test on Boiler No. 50 shows the performance of an internally fired boiler of special type, worked at a high rate of capacity. The rate of combustion was 17.3 pounds of coal per square foot of grate per hour, and this was obtained by the employment of a draught which reached the excessive pressure of 0.64 inches. Owing to the small amount of heating surface, the rate of evaporation was exceedingly high, being 5.2 pounds per square foot of heating surface per hour. This rate, however, was insufficient to work the boilers to their normal capacity, as based on the builders' rating of the power. The temperature of the escaping gases is above the standard of good economy, though not so high as might be expected in view of the conditions under which the boiler was operated. The evaporation per pound of combustible from and at 212 degrees, which is 10 pounds, is much below good boiler work, especially when considering the quality of the steam. The calorimeter showed that the steam contained 7 per cent. of moisture. On this test no deduction was made for the coke, ashes or soot which was deposited behind the bridge walls. Judging from the results of observations on a single boiler, such an allowance would increase the evaporative result about 4 per cent. The number of firemen employed for the six boilers was three, and they were not specially skilled in firing.

The capabilities of this type of boiler are best shown by a subsequent test, made on a single boiler of the plant, which was in the hands of an expert fireman. On this test the same class of fuel was used as before, and the coal, ashes and soot, deposited in the flue behind the bridge wall, were allowed for. The main results were as follows:

Coal per square foot of grate per hour,	21.3	lbs.
Percentage of ash,	7.0	per cent.
Water per square foot of heating surface per hour,	7.6	lbs.
Boiler pressure,	67.7	lbs.
Temperature of feed-water,	107.0	deg.
Temperature of escaping gases,	575.0	deg.
Draught suction,	0.57	in.
Percentage of moisture in steam,	0.5	per cent.
Water evaporated from and at 212 deg. per lb. of combustible,	11.06	lbs.

In view of the high temperature of the gases shown in this test, an evaporation of 11.06 pounds must be considered as excellent work. If the heating surface under these conditions had been increased so as to absorb the waste heat, as, for example, by means of a suitable flue heater, the evaporative result would readily have been brought up to the highest standard of economy.

Boiler No. 51.

Kind of boiler,	Vertical tubular (rolling pin).
Number used,	One.
Horse-power (basis 10 sq. ft.),	One hundred and forty.
Kind of coal,	Delaware and Lackawanna, Anthracite, Broken.
Age,	Ten years.

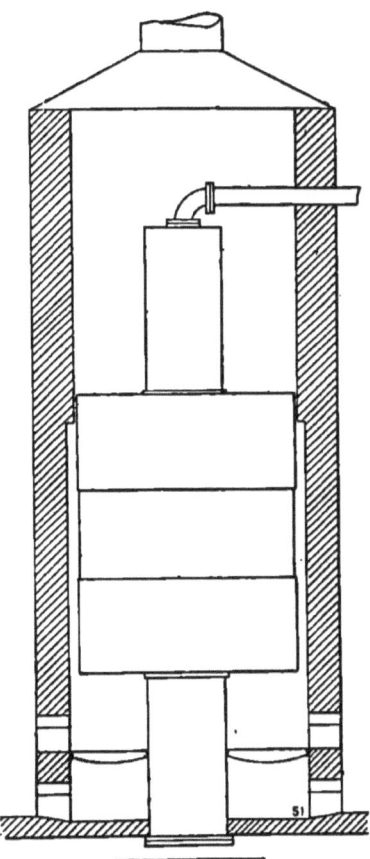

BOILER NO. 51, VERTICAL ELEVATION.

Boiler No. 51 is a vertical tubular boiler of the so-called "Rolling Pin" form, the principal features of which are shown in the following cut. In this boiler the main shell is completely filled with tubes, with the exception of the central space extending between the water leg and the dome. The grate surface forms an annular ring surrounding the water leg. The brick setting consists of a plain brick wall, surrounding the shell, and extending up to a height of about 10 feet above the upper tube sheet. This is surmounted by a wrought iron smoke stack. The water line is ordinarily carried to a point about 3 feet below the top of the main shell, and above this point the tubes are uncovered. A considerable portion of the heating surface is thus made steam heating surface, and the boiler produces steam which is in a more or less superheated condition. This form of vertical boiler has a proportion of 30 square feet of total heating surface to one of grate, and about two thirds of this is water-heating surface.

BOILER No. 51.

Dimensions of Boiler No. 51.

Diameter of main shell,	90 in.
Diameter of drum and water leg,	29 in.
Length of main shell and tubes,	9 ft. 9 in.
Number of tubes 2 inches outside diameter,	248
Area of water-heating surface,	941 sq. ft
Area of steam-heating surface,	426 sq. ft
Area of total heating surface,	1,367 sq. ft.
Area of grate surface,	45.6 sq. ft.
Area through tubes,	4.1 sq. ft
Width of air spaces and metal bars in grates,	1-2 in.
Area through stack,	8.7 sq. ft.
Height of stack,	60 ft.
Ratio of water-heating surface to grate surface,	20.6 to 1
Ratio of steam-heating surface to grate surface,	9.3 to 1
Ratio of total heating surface to grate surface,	30 to 1
Ratio of grate to tube area,	11 to 1

Results of Tests. Boiler No. 51.

	Test No.101.	Test No.102.	Test No.103.
Position of damper,	Wide open.	Partially closed.	Partially closed.
Manner of start and stop and kind of run,	Ordinary.	Ordinary.	Ordinary.
Duration, ... hrs.	11.7	11.2	11.7
Coal consumed (including wood equivalent), lbs.	9,161	5,705	5,897
Percentage of ash, . . per cent.	8.5	8.6	8.4
Water evaporated, . . . lbs.	61,355	44,597	48,735
Coal per hour, . . . lbs.	783	507.1	504
Coal per hour per square foot of grate, lbs.	17.1	11.1	11
Water per hour, . . . lbs.	5,244	3,964.2	4,165.4
Water per hour per square foot of water-heating surface, . . lbs.	5.6	4.3	3.6
Horse-power developed, . . H.P.	169.8	128.2	134.8
Boiler pressure, . . . lbs.	78	79	80
Temperature of feed-water, . deg.	132	132	134
Temperature of escaping gases, . deg.	600	480	434
Number of degrees of superheating, deg.	–	90	14
Water per pound of coal, . lbs.	6.70	7.82	8.26
Water per pound of coal from and at 212 degrees, ... lbs.	7.23	8.75	9.22
Water per pound of combustible from and at 212 degrees, . . . lbs.	8.18	9.56	10.07

The tests on Boiler No. 51 show the performance of this type of boiler under various conditions, when burning anthracite broken coal. Test No. 102 was made under the ordinary working conditions. Test No. 101 was made with damper wide open, so as to work the boiler to its maximum capacity. Test No. 103 was made under similar conditions to that of No. 102, with the single exception that the water line was carried to a higher point. The amount of water heating surface on this test was 1159 square feet, and that of steam heating surface 208 square feet, the ratios of these surfaces to the grate surface being respectively 25.4 and 4.6 to 1.

The evaporative results given in the Table take no account of the superheated condition of the steam. The amount of superheating in the case of Test No. 102 was 90 degrees. What allowance should be made for the superheat depends upon the character of the work which the steam is called upon to perform. The increased value of superheated steam is never less than the equivalent of the excess of the heat which it contains over saturated steam, whatever the use to which it is applied. For operating engines it has been found that the value of superheat is at least double that of heat expended in evaporation. If an allowance for superheating is made on Test No. 102 according to the heat added, the equivalent evaporative result per pound of combustible from and at 212 degrees is increased from 9.56 to 9.94 pounds. This result is not so high as that obtained from horizontal tubular boilers, working under favorable conditions. The temperature of the escaping gases on this test, which is 480 degrees, is excessive for anthracite coal, and it is evident that the inferior result is due, in a measure, at least, to the waste heat at this point.

The effect of increasing the water heating surface and reducing the amount of superheating is seen in the results of Test No. 103. The amount of superheating is only 14 degrees, and the temperature of the escaping gases is reduced to 434 degrees. The evaporative result is higher than that given on the low water test, being 10.07 from and at 212 degrees per pound of combustible, the increase amounting to about 5 per cent. Allowing for superheating as before according to the quantity of heat added, the results of the two tests are respectively 9.94 pounds and 10.15 pounds, and the difference between them is about 2 per cent. The increase in the water heating surface enabled the boiler to utilize a larger proportion of the heat of combustion of the coal to the extent of about 2 per cent.

A comparison of Tests No. 101 and No. 102 shows the unfavorable effect produced upon the vertical type of boiler by a high rate of combustion. The quantity of coal burned in a given time in Test No. 101 was 54 per cent. larger than in Test No. 102, and the evaporation per pound of coal was reduced 14.4 per cent. This reduction is accounted for by the increased temperature of the escaping gases, which was from 480 degrees to 600 degrees. The amount of power developed on the capacity test was large, and the ease here shown with which the vertical boiler generates steam should not be lost sight of. With a stack only 60 feet high, the boilers developed 21 per cent. more than the rated capacity (without allowance for the greater efficiency of superheated steam), and the rate of combustion was 17.1 pounds of coal per square foot of grate per hour.

Boiler No. 52.

Kind of boiler,	Vertical tubular with fire-box.
Number used,	One.
Horse-power (basis 10 sq. ft.),	Twenty.
Kind of coal,	Anthracite, Lehigh, egg.
Age,	Four months.

Boiler No. 52 is of the vertical type, arranged in the manner shown in the following cut. It is supported by a brick ash pit, but otherwise it has no brick setting, the exterior surface being covered with a non-conducting cement. The steam, before leaving the boiler, passes through a chamber in the interior of the shell. The top and bottom of this chamber are connected by means of tubes which surround the respective fire tubes of the boiler, the annular space between the two tubes

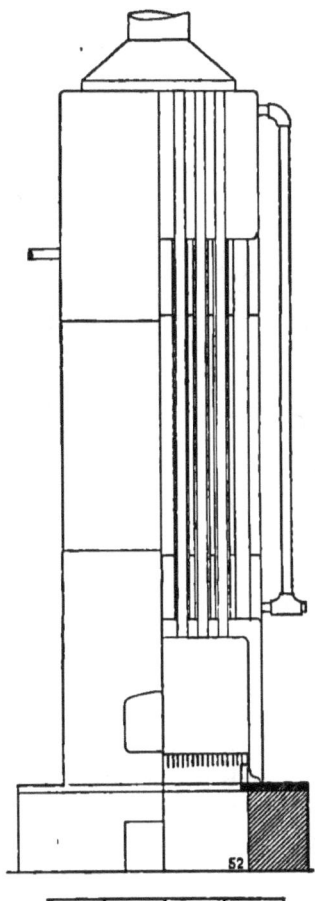

BOILER No. 52, VERTICAL ELE-
VATION AND SECTION.

being filled with water. When the boiler is in operation, there is a circulation of water from the bottom of the boiler upward through these annular spaces, to the top of the boiler, and thence downward through an outer row of tubes, which are provided for the return of the water. The upper ends of the fire tubes are bare, and the steam, on leaving the steam space, is by this means superheated. Owing to the subsequent passage of the steam through the drum, the superheat is absorbed, and the steam leaves the boiler in a saturated state. The boiler is surmounted by a smoke stack, which is so placed that it interfered with the cleaning of the tubes, and these had not been brushed out since the boiler was started.

Dimensions of Boiler No. 52.

Diameter of shell,	40 in.
Diameter of fire-box and grate,	33 in.
Height between heads and length of tubes,	9 ft. 6 in.
Number of tubes 2 inches outside diameter,	42
Height of fire-box,	24 in.
Area of heating surface (total),	205 sq. ft.
Area of grate surface,	6.1 sq. ft.

BOILER No. 52.

Area through tubes,	0.7 sq. ft.
Area through stack,	0.78 sq. ft.
Height of stack,	52 ft.
Ratio of total heating surface to grate surface,	33.5 to 1
Ratio of grate surface to tube area,	8.7 to 1

Results of Tests, Boiler No. 52.

	Test No. 104.	Test No. 105.
Manner of start and stop,	Ordinary with preliminary heating.	Ordinary with preliminary heating.
Kind of run,	Continuous.	Continuous.
Duration, hrs.	7.25	7.25
Coal consumed (including wood equivalent), lbs.	664	404
Percentage of ash, per cent.	8.9	9.4
Water evaporated, lbs.	5,676	3,597
Coal per hour, lbs.	91.6	55.8
Coal per hour per square foot of grate, lbs.	15	9.1
Water per hour, lbs.	782.9	496.2
Water per hour per square foot of total heating surface, lbs.	3.8	2.42
Horse-power developed, H. P.	24.5	15.5
Boiler pressure, lbs.	89	91
Temperature of feed-water, deg.	163	182
Temperature of escaping gases, deg.	596	446
Water per pound of coal, lbs.	8.55	8.90
Water per pound of coal from and at 212 degrees, lbs.	9.30	9.52
Water per pound of combustible from and at 212 degrees, lbs.	10.22	10.51

The tests on Boiler No. 52 were made with two rates of combustion, viz., 15 pounds and 9.1 pounds per square foot of grate per hour. The high rate of combustion in Test No. 104 is noticeable, considering the low height of the stack, which was 52 feet, and this was not the maximum, as the damper was partially closed. The arrangement of tubes in the vertical type of boiler, with stack directly above, is favorable for the production of a large capacity with a feeble draught, and this fact is here exemplified. The results of these tests furnish evidence that the use of a fire box in vertical boilers is advantageous. Comparing Test No. 105 on this boiler, with Test No. 102 made on vertical boiler No. 51 which is set in brick work, there is a noticeable improvement in favor of the firebox boiler, the two quantities being respectively 10.51 pounds and 9.94 pounds, allowance here being made for the superheat. Notwithstanding the small size of the fire-box boiler, and the foul condition of the tubes, it secured much the better result.

A comparison of the two tests on this boiler shows unmistakably that under equally favorable conditions, a rapid combustion is more economical than a slow combustion. The evaporation on test No. 104 with rapid combustion is less than 3 per cent. below that on the other test, while the temperature of the gases is 150 degrees higher. If the waste heat of the gases had been utilized by the employment of a heater, or by otherwise increasing the heating surface, the performance would have been improved (judging from the results of heater tests elsewhere given) at least 10 per cent., thus leaving 7 per cent. or more net improvement due to more rapid combustion.

Boiler No. 53.

Kind of boiler,	Vertical tubular.
Number used,	One.
Horse-power (basis 10 sq. ft.),	Seventy-five.
Age,	New.

Boiler No. 53 is a vertical tubular boiler of the same general form as that shown in the cut of Boiler No. 52. Instead of being provided, however, with a fire box, as in the one referred to, it has a brick furnace, and furthermore the steam is discharged from the boiler immediately after its generation, without passing through the interior chamber. The exterior surface of the shell is protected with a somewhat inefficient covering of cement. The boiler is provided with a forced draught, which is produced by a blower discharging under the ash pit. For the purpose of burning screenings on test No. 108 a plate was employed in place of the ordinary grate, and this was perforated with one half inch holes, located $1\frac{1}{2}$ inches from center to center. The tubes of the boiler were not cleaned during the test, and had not been cleaned since the boiler was started two weeks before.

BOILER No. 53.

Dimensions of Boiler No. 53.

Diameter of shell,	64 in.
Height of shell and length of tubes,	14 ft.
Number of tubes, 2 1-2 inches outside diameter,	88
Diameter of brick furnace,	54 in.
Area of water-heating surface,	513 sq. ft.
Area of steam-heating surface,	248 sq. ft.
Area of grate surface,	15.9 sq. ft.
Area through tubes,	2.4 sq. ft.
Width of air spaces and metal bars in grates,	Air, 5-8 in., metal, 3-4 in.
Area through stack,	5.9 sq. ft.
Height of stack,	60 ft.
Distance from grate to tube sheet,	4 ft.
Ratio of water-heating surface to grate surface,	32.3 to 1
Ratio of steam-heating surface to grate surface,	15.6 to 1
Ratio of grate to tube area,	6.5 to 1

Results of Tests, Boiler No. 53.

	Test No. 106.	Test No. 107.	Test No. 108.
Kind of coal,	Anthracite Schuylkill, Broken.	Anthracite Schuylkill, Broken.	Anthracite Schuylkill, Screenings.
Draught,	Natural.	Forced.	Forced.
Manner of start and stop,	Ordinary with preliminary heating.	Ordinary with preliminary heating.	Ordinary with preliminary heating.
Kind of run,	Continuous.	Continuous.	Continuous.
Duration, . . . hrs.	10	10.7	7.7
Coal consumed, dry (including wood equivalent), lbs.	2,452	4,134	3,010
Percentage of ash, per cent.	11.6	13.6	13.4
Water evaporated, lbs.	19,583	29,846	19,937
Coal per hour, lbs.	245.2	384.5	392.4
Coal per hour per square foot of grate, lbs.	15.4	24.1	24.7
Water per hour, lbs.	1,958.3	2,776.4	2,599.4
Water per hour per square foot of water-heating surface, lbs.	3.8	5.4	5.1
Horse-power developed, H. P.	64.3	91.2	85.4
Boiler pressure, lbs.	61	73.2	73.1
Temperature of feed-water, deg.	125	125	124
Temperature of escaping gases, deg.	478	573	505
Number of degrees of superheating, deg.	42	59	51
Water per pound of coal, lbs.	7.98	7.21	6.62
Water per pound of coal from and at 212 degrees, lbs.	8.95	8.10	7.45
Water per pound of combustible from and at 212 degrees, lbs.	10.13	9.38	8.61

The tests on Boiler No. 53 were made to determine the economy of the boiler when burning anthracite coal, both with natural and forced draught, and that obtained when using athracite screenings with forced draught. Comparing test No. 107 with No. 106, the use of the blower increased the power developed by the boiler from 64.3 horse power to 91.2 horse power, or 42 per cent., and this was accompanied by a reduction in the evaporation per pound of combustible of 7.5 per cent. After allowing for the small difference in the amount of superheating, the reduction becomes 6.7 per cent. The effect of the higher rate of combustion was to increase the temperature of the escaping gases from 478 to 573 degrees, and to this large increase in the amount of waste heat may be attributed the reduction in economy noted. Comparing these results with those obtained from boilers having a less proportion of water heating surface, it is seen that this boiler has the advantage. In Boiler No. 51 for example, the ratio of water heating surface to grate surface is 20.6 to 1, and the evaporation per pound of combustible from and at 212 degrees, corrected for superheating, is 9.94 pounds. In Boiler No. 53, under consideration, the ratio of water heating surface to grate surface is 32.3 to 1, and the result obtained, corrected for superheating, is 10.34 pounds of water per pound of combustible. The result obtained by the use of screenings with forced draught compares favorably with that obtained under the same circumstances with the standard coal, being inferior to the coal result only 8.1 per cent. No allowance is made in the tests with forced draught for the cost of power required to operate the blower.

Boiler No 54.

Kind of boiler,	Vertical tubular (rolling pin).
Number used,	One.
Horse-power (basis 10 square feet),	One hundred and forty.
Kind of coal,	Two parts Anthracite Screenings, one part Cumberland.
Age,	Ten years.

Boiler No. 54 is a vertical tubular boiler arranged and set in the manner shown in the following cut. The style of setting differs from the ordinary practice, the sides of the furnace being provided with perforated tiles, through which air is admitted above the fuel. The air is first passed up and down through ducts in the walls, before it is discharged into the furnace.

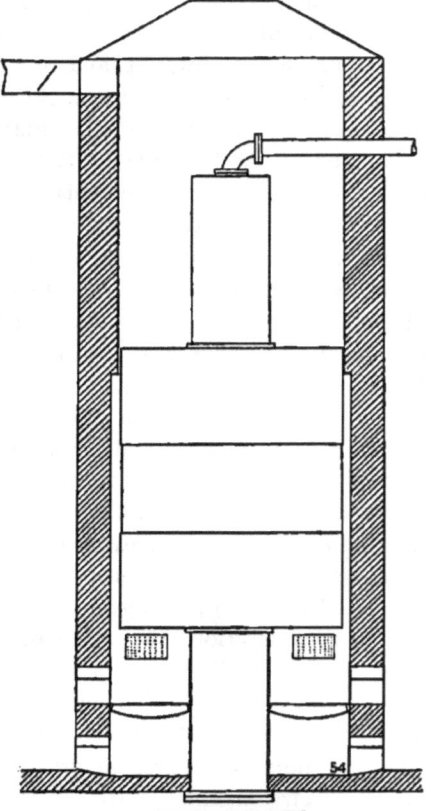

BOILER No. 54, VERTICAL ELEVATION.

Dimensions of Boiler No. 54.

Diameter of main shell,	90 in.
Diameter of drum and water leg,	29 in.
Length of main shell and tubes,	9 ft. 9 in.
Number of tubes, 2 inches outside diameter,	248
Area of water-heating surface,	936 sq. ft.
Area of steam-heating surface,	517 sq. ft.
Area of total heating surface,	1,453 sq. ft.
Area of grate surface,	45.6 sq. ft.
Area through tubes,	4.1 sq. ft.
Area through flue,	2.8 sq. ft.
Height of chimney,	125 ft.
Width of air spaces and metal bars in grates,	1-2 in.

Ratio of water-heating surface to grate surface, . 21.1 to 1
Ratio of steam-heating surface to grate surface, . 10.7 to 1
Ratio of total heating surface to grate surface, . 31.8 to 1
Ratio of grate to tube area, ... 11 to 1
Ratio of grate to flue area, ... 16.1 to 1

Results of Tests, Boiler No. 54.

	Test No. 109.	Test No. 110.	Test No. 111.	Test No. 112.
Kind of coal,	Mixture 2 parts Anthracite Screenings, 1 part Cumberland.		George's Creek Cumberland	Anthracite Lehigh. Broken.
Conditions as to admission of air over fuel,	Air admitted.	Air excluded.	Air excluded.	Air excluded.
Manner of start and stop and kind of run,	Ordinary.	Ordinary.	Ordinary.	Ordinary.
Duration, . . hrs.	11	11.7	11.7	12.5
Coal consumed, dry (including wood equivalent), . . lbs.	5,399	6,211	4,070.4	5,592
Percentage of ash, per cent.	18.8	16.7	8.2	15.1
Water evaporated, lbs.	33,210	40,001	31,888	37,504
Coal per hour, . lbs.	483.2	528.6	346.5	447.4
Coal per hour per square foot of grate, . lbs.	10.7	11.6	7.3	9.8
Water per hour, . lbs.	3,019.1	3,404.3	2,713.8	3,000.3
Water per hour per square foot of water-heating surface, . . lbs.	3.2	3.6	2.9	3.2
Horse-power developed, H. P.	104.6	118.1	94	103.8
Boiler pressure, . lbs.	72.5	72.5	72.7	72.2
Temperature of feed-water, . . deg.	42	42	44	44
Temperature of escaping gases, . . deg.	443	434	413	449
Number of degrees of superheating, . deg.	93	93	73	89
Draught suction, . in.	0.26	0.32	0.07	0.05
Number of firings, .	22	31	23	15
Number of times slice bar or stoker used,	14	22	22	3
Water per pound of coal, lbs.	6.15	6.44	7.83	6.71
Water per pound of coal from and at 212 degrees, . . lbs.	7.46	7.79	9.46	8.11
Water per pound of combustible from and at 212 degrees, . lbs.	9.06	9.27	10.27	9.54

NOTE.—The coal when fired contained 10 per cent. of moisture.

The first two tests on Boiler No. 54, viz., tests No. 109 and No. 110, had for an object the determination of the effect which the admission of air above the fuel has upon the economy and capacity of the boiler. The damper was kept in a constant position for both tests, and there was a steady draught of 0.7 inch on the chimney side of the damper. The effective draught in the boiler flue was 0.26 inch when the air was admitted and 0.32 inch when it was excluded, and, as a consequence, the power developed was increased, by excluding the air, from 209.2 horse power to 236.2 horse power, an increase of 13 per cent. In the matter of economy, the exclusion of air was followed by an increase of 4.3 per cent. in the evaporation per pound of coal, and of 2.3 per cent. per pound of combustible.

Test No. 110 taken in connection with tests No. 111 and No. 112, form a series of tests which show the relative economy of different kinds of fuel. On each of these three tests the air ducts remained closed. The evaporation per pound of Cumberland coal was 16.6 per cent. greater, and that per pound of mixed fuel 4 per cent. less, than the performance with anthracite coal. The cost of fuel required to evaporate 30,000 pounds of water from and at 212 degrees, at the prices which ruled at the time of the tests, is as follows:

	Cumberland.	Anthracite Broken.	Mixture.
Cost per ton of 2,200 pounds,	$5 50	$5 50	$3 50
Cost for 30,000 pounds of steam,	7 81	9 24	6 03

From this it appears that the cost of Cumberland coal required to do a given amount of work was 15.5 per cent. less, and that of the mixture 34.7 per cent. less, than the cost of anthracite coal.

The relatively small amount of draught required in burning the standard grades of coal is noticeable, although the boiler developed somewhat less than the rated power, and the rate of combustion was low. The draught on test No. 112 is less than one-sixth of the full capacity of a good chimney. The Cumberland coal required a slightly stronger draught than the anthracite coal, and the mixed fuel required six times as much.

The relative labor of firing the various coals is seen in the number of times of firing and using the slicing bar. The mixed fuel is at the greatest disadvantage in this respect.

The economical results, as a whole, must take into account the character of the steam, which in the various cases was superheated from 73 degrees to 93 degrees. Making the most favorable allowance for the increased value of the steam in this condition, the results are much below that obtained from the best types of boilers. The small proportion of water heating surface, the consequent rapid evaporation per square foot of that surface, and the resulting high temperature of the escaping gases, are characteristic of this type of boiler, and account for the unfavorable character of the results.

Boiler No. 55.

Kind of boiler,	Vertical tubular.
Number used,	Five.
Horse-power (collective, basis 10 sq. ft.),	Four hundred and seventy-five.
Age,	One year.

Dimensions of Boiler No. 55.

Diameter of shell,	64 in.
Length between heads and length of tubes,	14 ft.
Number of tubes (collective), 3 inches outside diameter,	420
Area of water-heating surface,	3,300 sq. ft.
Area of steam-heating surface,	1,455 sq. ft.
Area of total heating surface,	4,755 sq. ft.
Area of grate surface,	157.5 sq. ft.
Area through tubes,	17.2 sq. ft.
Area through flue,	27.1 sq. ft.
Distance of grate to tube sheet,	8 ft. 7 in.
Ratio of water-heating surface to grate surface,	29.0 to 1
Ratio of steam-heating surface to grate surface,	9.2 to 1
Ratio of grate to tube area,	9.1 to 1

BOILER No. 55.

Results of Tests, Boiler No. 55.

	Test No. 113.	Test No. 114.	Test No. 115.	Test No. 116.
Number of boilers,	Five.	Five.	One.	One.
Kind of coal,	Anthracite Lehigh Broken.	Anthracite Lehigh Broken.	Anthracite Lehigh Broken.	Mixture 3 parts screenings, 1 part Cumberland coal.
Manner of start and stop and kind of run,	Ordinary.	Ordinary.	Ordinary.	Ordinary.
Duration, . hrs.	11.5	11.5	12	12.5
Coal consumed (including wood equivalent), lbs.	25,904	22,925	3,527	3,614
Percentage of ash, per cent.	10.3	13.2	13.2	16.5
Water evaporated, lbs.	158,705	148,686	22,599	22,017
Coal per hour, . lbs.	2,252	1,990	293.9	289.1
Coal per hour per square foot of grate, . lbs.	14.3	12.7	9.3	9.2
Water per hour, . lbs.	13,800.4	12,929.2	1,883.2	1,761.4
Water per hour per square foot of water-heating surface, . lbs.	4.2	3.9	2.8	2.8
Horse-power developed, H. P.	555.6	522.5	65.4	61.4
Boiler pressure, . lbs.	72.1	72	72.1	70.2
Temperature of feedwater, . deg.	126	121	39	39
Temperature of escaping gases, . deg.	545	509	417	462
Number of degrees of superheating, . deg.	80	73	71	70
Draught suction, . in.	0.16	0.16	0.09	0.03
Water per pound of coal, lbs.	6.13	6.49	6.41	6.09
Water per pound of coal from and at 212 degrees, . lbs.	6.87	7.33	7.77	7.38
Water per pound of combustible from and at 212 degrees, . lbs.	7.63	8.43	8.96	8.85

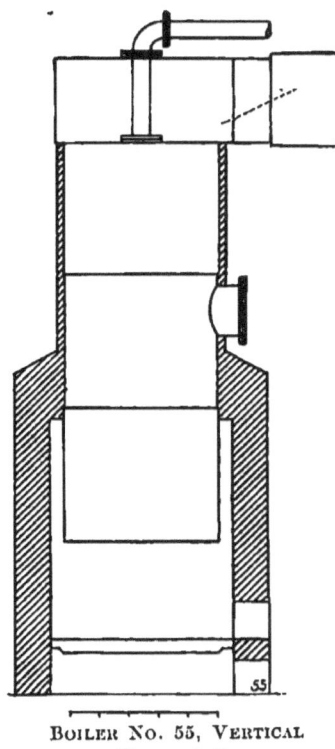

BOILER No. 55, VERTICAL ELEVATION.

Boiler No. 55 consists of a plant of five vertical tubular boilers of the type shown in the following cut. This boiler consists of a plain vertical shell, containing tubes placed in two sections, arranged with an open space between them for purposes of inspection and cleaning. It is set with a brick furnace, the walls of which extend about half way up on the sides of the boiler. Above this point the shell is covered with non-conducting cement. The boilers are fitted with tubes having an external diameter of 3 inches, which is somewhat larger than those ordinarily used for vertical boilers.

Tests No. 113 and No. 114 were made on the full plant of five boilers, while tests No. 115 and No. 116 were made on a single boiler of the plant. The first three tests show the performance of the boilers with anthracite coal, working under three different rates of combustion, the highest being 14.3 pounds per square foot of grate per hour, and the lowest 9.3 pounds. There is a continuous improvement in the evaporative result as the rate is decreased, though the best result obtained cannot be looked upon as favorable. The highest performance is 8.96 pounds of water from and at 212 degrees per pound of combustible, with 71 degrees of superheating. The low degree of economy which was realized may be attributed in part to the high temperature of the escaping gases, and in part to loss produced by radiation from the brick furnace, and by leakage of air through the brick work. The large area through the tubes doubtless contributed in some degree to the unfavorable result.

The low draught required for a vertical boiler is here noticeable, 16 inches being sufficient to burn 14.3 pounds of anthracite coal per square foot of grate per hour. In the case of the test with the mixed fuel the draught suction was .03 inch.

Comparing the results of test No. 116, made with mixed fuel, with that of No. 115, made with anthracite coal, the evaporation with the mixture, based on coal, is 5 per cent. less, and based on combustible, 1 per cent. less, than that with the standard fuel. Taking into account the cost of these fuels at the time of the tests, the economy in favor of the mixed fuel amounts to 33 per cent.

Boiler No. 56.

Kind of boiler,	Vertical tubular (nest).
Number used,	Two.
Horse-power (collective, basis 10 sq. ft.),	Three hundred and eighty.
Kind of coal,	Anthracite chestnut No. 2.
Age,	Ten years.

Dimensions of Boiler No. 56.

Diameter of central shell,	36	in.
Length of central shell,	21	ft.
Diameter of outer shells (6 in number, each),	30	in.
Length of outer shells and length of tubes,	10	ft.
Number of tubes (collective), 2 inches outside diameter,	516	
Area of water-heating surface,	2,224	sq. ft.
Area of steam-heating surface,	1,566	sq. ft.
Area of total heating surface,	3,790	sq. ft.
Area of grate surface,	143	sq. ft.
Area through tubes,	8.6	sq. ft.
Area through flue,	10	sq. ft.
Ratio of water-heating surface to grate surface,	15.5	to 1
Ratio of steam-heating surface to grate surface,	11	to 1
Ratio of grate surface to tube area,	16.6	to 1

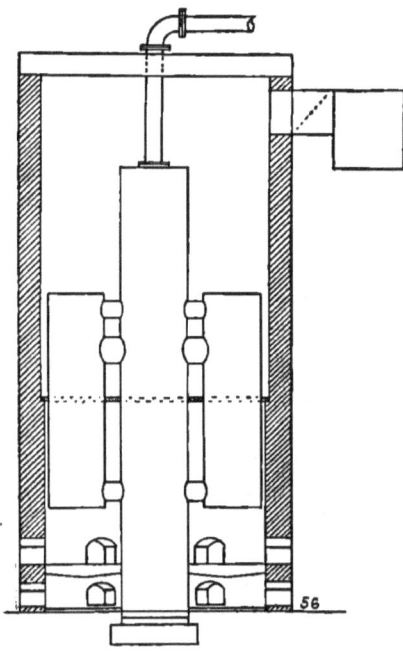

BOILER NO. 56, VERTICAL ELEVATION.

Boiler No. 56 is a vertical tubular boiler, consisting of six vertical shells filled with tubes, surrounding and communicating with a central shell, as shown in the following cut.

The whole is enclosed in a circular wall which forms the setting. The flue space is separated from the furnace space by iron plates fitting around the shells. This type of boiler does not differ in principle from the ordinary vertical boiler, but it differs somewhat in the proportions. It has a low proportion of water-heating surface to grate surface, namely, 15.5 to 1, and a correspondingly high proportion of steam-heating surface. The tube area is small.

Results of Test, Boiler No. 56.

		Test No. 117.
Manner of start and stop and kind of run,		Ordinary.
Duration,	11	hrs.
Coal consumed,	11,862	lbs.
Percentage of ash,	12.9	per cent.
Water evaporated,	82,584	lbs.
Coal per hour,	1,078.4	lbs.
Coal per hour per square foot of grate,	7.5	lbs.
Water per hour,	7,507.6	lbs.
Water per hour per square foot of water-heating surface,	3.4	lbs.
Horse-power developed,	242.8	H. P.
Boiler-pressure,	83	lbs.
Temperature of feed-water,	120	deg.
Temperature of escaping gases,	468	deg.
Number of degrees of superheating,	31	deg.

Draught suction,	0.08	in.
Water per pound of coal,	6.96	lbs.
Water per pound of coal from and at 212 degrees,	7.86	lbs.
Water per pound of combustible from and at 212 degrees,	8.91	lbs.

The test on Boiler No. 56 shows the performance of a vertical boiler, which was deficient in heating surface, when using one of the small grades of anthracite coal. The rate of combustion is low, and there is consequently a small amount of power developed. The power is nearly 40 per cent. below the nominal capacity. In spite of this low rate the temperature of the escaping gases is high, being 468 degrees, and the economic result is inferior, not only to the best class of boilers, but also to the best performance of this particular class. In this boiler, as in other vertical boilers set in brick work, there is a large exterior surface exposed to the air. This surface not only causes radiation, but owing to the unstable condition of brick-work which has had a long period of service, it provides many avenues for the entrance of air both to the furnace and to the flue space. The effect of unnecessary admission of air is to reduce the efficiency of the fuel and to lower the temperature of the gases. The last-named effect causes the apparent indication of waste heat to be much less than the real quantity.

Boiler No. 57.

Kind of boiler,	Vertical tubular (rolling pin).
Number used,	One.
Horse-power (basis 10 square feet),	One hundred and forty.
Kind of coal,	Anthracite Lackawanna, egg.
Age,	Several years.

Boiler No. 57 is a vertical boiler of the rolling pin type, having the general features shown in the cut of Boiler No. 51.

Dimensions of Boiler No. 57.

Diameter of main shell,	90	in.
Diameter of drum and water leg,	31	in.
Length of main shell and tubes,	10	ft.
Number of tubes, 2 inches outside diameter,	248	
Area of water-heating surface,	897	sq. ft.
Area of steam-heating surface,	505	sq. ft.
Area of total heating surface,	1,402	sq. ft.

Area of grate surface, ... 45 sq. ft.
Area through tubes, ... 4.1 sq. ft.
Height of chimney, ... 90 ft.
Width of air spaces and metal bars in grates, . Air 3-8 in., metal 1-2 in.
Distance of grate to tube sheet, ... 2 ft. 8 in.
Ratio of water-heating surface to grate surface, . . 19.9 to 1
Ratio of steam-heating surface to grate surface, . . . 11.3 to 1
Ratio of grate surface to tube area, ... 10.9 to 1

Results of Tests, Boiler No. 57.

	Test No. 118.	Test No. 119.
Manner of start and stop and kind of run,	Ordinary.	Ordinary.
Duration, ... hrs.	10.37	10.12
Coal consumed (including wood equivalent), ... lbs.	5,602	6,472
Percentage of ash, . . . per cent.	17	17.1
Water evaporated, ... lbs.	35,482	39,646
Coal per hour . . . . . lbs.	540.4	637.7
Coal per hour per square foot of grate, lbs.	12	14.2
Water per hour, ... lbs.	3,422.7	3,915.7
Water per hour per square foot of water-heating surface, ... lbs.	3.8	4.4
Horse power developed, . . H. P.	119.8	137
Boiler pressure, ... lbs.	79.5	77.8
Temperature of feed-water, . . deg.	52	49
Temperature of escaping gases, . deg.	520	–
Draught suction, ... in.	0.12	0.24
Number of degrees of super-heating, deg.	65	89
Water per pound of coal, . . . lbs.	6.33	6.13
Water per pound of coal from and at 212 degrees, ... lbs.	7.61	7.37
Water per pound of combustible from and at 212 degrees, ... lbs.	9.17	8.87

The tests on Boiler No. 57 show the performance of a vertical tubular boiler under two rates of combustion, both of which are comparatively high for the small proportion of heating surface to grate surface which is here found, viz., 19.9 to 1. As in all boilers of this class, the temperature of the escaping gases is high, and there is a correspondingly low economic result, There is a small reduction in the evaporation produced by the high rate of combustion. It is noticeable in the comparison of the two tests that it required double the draught suction to increase the rate of combustion from 12 pounds to 14.2 pounds per square foot of grate per hour.

Boiler No. 58.

Kind of boiler,	Vertical tubular.
Number used,	Two.
Horse-power (collective, basis 12 sq. ft.),	Two hundred and ten.
Kind of coal,	Clearfield bituminous.
Age,	Four years.

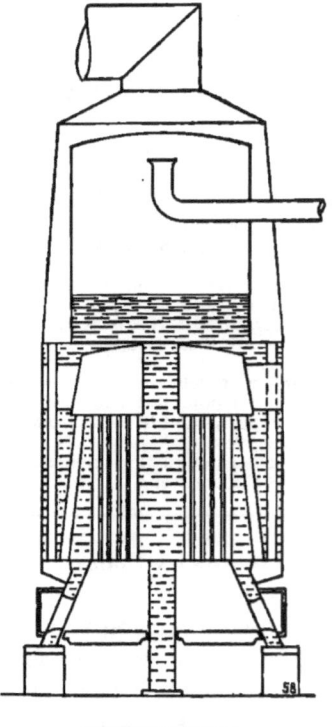

BOILER NO. 58, VERTICAL SECTION.

Boiler No. 58 is a vertical boiler of somewhat novel form, the general features of which are shown in elevation in the following cut. The boiler is of the fire box type, the water leg of which is provided with two opposite openings for the introduction of fuel. The tubes are divided into three sections arranged in concentric circles. The first section embraces the inner tubes. These carry the products of combustion upward from the furnace to a combustion chamber, which is located inside the boiler, below the water line. The second embraces those in the next circle, and through these the products of combustion pass downward. The third section embraces the outer tubes, which carry the gases upward to the space surrounding the steam dome, whence they finally enter the chimney flue. The proportion of water heating surface to grate surface, which is 35.1 to 1, is nearly double that which is found in the rolling pin type of boiler, and the proportion of steam heating surface to grate surface is less than one-half of that found in the boiler noted. The arrangement of tubes, which provides for a downward current of the gases through a part of their passage, secures a distribution of the heating surface whereby it is made efficient throughout its whole extent.

Dimensions of Boiler No. 58.

Diameter of shell 7½ feet high,	8 ft.
Diameter of fire-box,	6.5 ft.
Number of tubes in first section (collective), 3 inches outside diameter,	280
Number of tubes in second section (collective), 4 inches outside diameter,	76
Number of tubes in third section (collective), 4½ inches outside diameter,	76
Length of tubes in first and second sections,	5 ft.
Length of tubes in third section,	7.5 ft.
Diameter of dome,	6 ft.
Height of dome,	7.5 ft.
Area of water-heating surface,	2,273 sq. ft.
Area of steam-heating surface,	263 sq. ft.
Area of total heating surface,	2,536 sq. ft.
Area of grate surface,	64.8 sq. ft.
Area through flue,	12 sq. ft.
Width of air spaces and metal bars in grates,	Air 1-2 in., metal 3-8 in.
Ratio of water-heating surface to grate surface,	35.1 to 1
Ratio of steam-heating surface to grate surface,	4 to 1
Ratio of grate surface to smallest tube area,	12.1 to 1

Results of Test, Boiler No. 58.

Test No. 120.

Manner of start and stop and kind of run,		Ordinary.
Duration,	11.5	hrs.
Coal consumed, dry (including wood equivalent),	7,682	lbs.
Percentage of ash,	9.3	per cent.
Water evaporated,	72,207	lbs.
Coal per hour,	668	lbs.
Coal per hour per square foot of grate,	10.3	lbs.
Water per hour,	6,278.9	lbs.
Water per hour per square foot of water-heating surface,	2.8	lbs.
Horse-power developed,	213.2	H. P.
Boiler pressure,	76.5	lbs.
Temperature of feed-water,	66	deg.
Temperature of escaping gases,	423	deg.
Water per pound of coal,	9.4	lbs.
Water per pound of combustible from and at 212 degrees,	11.19	lbs.
Water per pound of coal from and at 212 degrees,	12.29	lbs.

The test on Boiler No. 58 exhibits the performance of a vertical tubular boiler, so planned as to overcome the losses that attend the work of vertical boilers as ordinarily set in brick work. Unnecessary air leakage into the furnace is prevented by the use of a fire box. The ratio of heating surface to grate surface is as large as that found in many efficient horizontal tubular boilers. The rate of combustion is moderately high, and the temperature of the escaping gases is not excessive for bituminous coal. All these conditions are favorable to economy, and they go hand-in-hand with the high character of the evaporative result which the test gives. An evaporation of 12.29 pounds of water from and at 212 per pound of combustible, which was obtained in this case, is seldom exceeded by any type of boiler.

Boiler No. 59.

Kind of boiler,	Vertical tubular (rolling pin).
Number used,	Two.
Horse-power (collective, basis 10 sq. ft.),	Two hundred and eighty.
Kind of coal,	George's Creek Cumberland.
Age,	Twelve years.

Boiler No. 59 embraces a plant of two vertical boilers of the rolling pin type, of the general form shown in elevation in the cut of Boiler No. 51. They are provided with a flue heater, the location of which, with reference to the boilers, is shown in ground plan in the following cut. The heater consists of vertical cast-iron pipes, similar to that of Boiler No. 33, the exposed surface of which has nearly as large an area as that of the total water heating surface of the boilers.

Dimensions of Boiler No. 59.

Diameter of main shell,	90 in.
Diameter of drum and water leg,	30 in.
Length of main shell and tubes,	10 ft.
Number of tubes (collective), 2 inches outside diameter,	496
Area of water-heating surface,	1,880 sq. ft.
Area of steam-heating surface,	924 sq. ft.
Area of grate surface,	90 sq. ft.
Area through tubes,	8.2 sq. ft.
Width of air spaces and metal bars in grates,	Air 1-2 in., metal 5-8 in.
Height of chimney,	80 ft.
Ratio of water-heating surface to grate surface,	20.9 to 1
Ratio of steam-heating surface to grate surface,	10.3 to 1
Ratio of grate surface to tube area,	11 to 1
Area of heating surface in flue heater,	1,600 sq. ft.

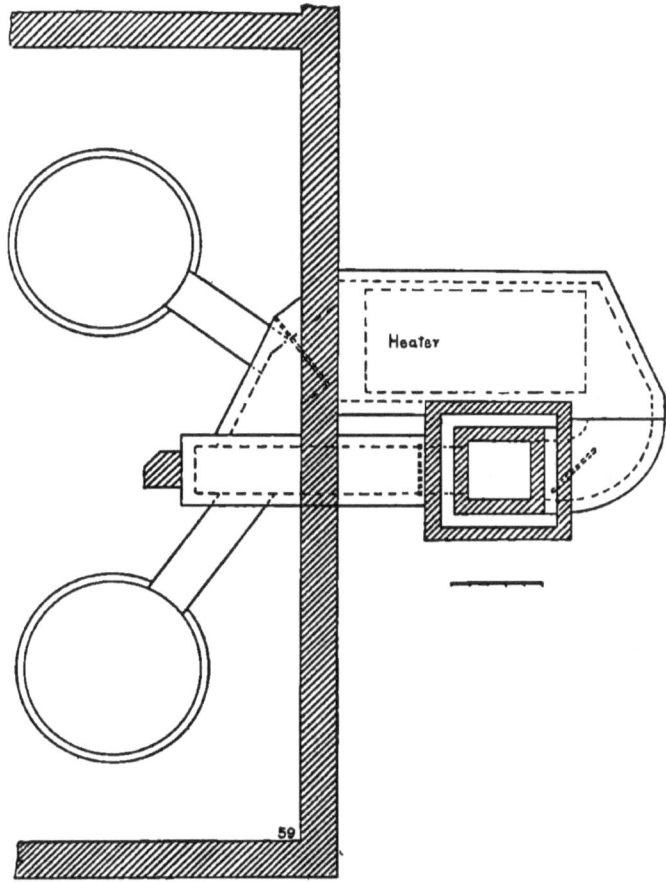

BOILER No. 59, GROUND PLAN SHOWING LOCATION OF BOILERS AND FLUE HEATER.

BOILER No. 59.

Results of Tests, Boiler No. 59.

	Test No. 121.	Test No. 122.
	Heater in use.	Heater not in use.
Manner of start and stop and kind of run,	Ordinary.	Ordinary.
Duration, ... hrs.	11.5	11.5
Coal consumed, dry (including wood equivalent), ... lbs.	7,856	10,282
Percentage of ash, . . . per cent.	8	8.4
Water evaporated, ... lbs.	72,002	72,959
Coal per hour, ... lbs.	683.1	894.1
Coal per hour per square foot of grate, lbs.	7.6	9.9
Water per hour, ... lbs.	6,261	6,344.3
Water per hour per square foot of water-heating surface, ... P.	3.3	3.4
Horse-power developed, . . Hlbs.	210.3	213.6
Boiler pressure, ... lbs.	58	59
Temperature of feed-water entering heater, ... deg.	88	–
Temperature of feed-water entering boiler, ... deg	225	85
Temperature of escaping gases leaving boiler, ...deg.	618	645
Temperature of escaping gases leaving heater, ... deg.	365	–
Number of degrees of superheating, deg.	50	52
Draught suction, . . . in.	0.09	–
Water per pound of coal, . . lbs.	9.16	7.10
Water per pound of coal from and at 212 degrees, ... lbs.	–	8.23
Water per pound of combustible from and at 212 degrees, . . . lbs.	–	8.09

NOTE.—The coal when fired contained 3 per cent. of moisture.

The tests on Boiler No. 59 had for a principal object the determination of the economy secured by the use of a flue heater in connection with vertical boilers. They are of special interest in view of the large amount of waste heat which ordinarily escapes to the flue in this type of boiler. In test No. 121 the heater was in operation, while in test No. 122 the heater was shut off, and the gases passed directly to the chimney. The employment of the heater reduced the temperature of the escaping gases 618° — 365° = 253 degrees, and increased the temperature of the feed water 225° — 88° = 137 degrees, resulting in an increase of the evaporation per pound of coal of 7.10 to 9.16 pounds or 29 per cent. When the heater was not in use, the economy secured was noticeably low, which could not be otherwise with the high flue temperature which existed, viz., 645 degrees.

Considering the results of test No. 121 as applied to the plant as a whole, the evaporation per pound of combustible from and at 212 becomes 11.54 pounds. Allowing for the superheating, which amounted to 52 degrees, the result is equivalent to 11.84 pounds of water evaporated. This is not far below the best results commonly obtained from well proportioned horizontal tubular boilers not provided with a flue heater.

Boiler No. 60.

Kind of boiler,	Vertical tubular (fire box).
Number used,	Two.
Horse-power (collective, basis 13 sq. ft.),	Two hundred and sixty.
Kind of coal,	George's Creek Cumberland.
Age,	Sixteen months.

Dimensions of Boiler No. 60.

Diameter of shell,	60 in.
Height between heads and length of tubes,	15 ft.
Number of tubes (collective), 2½ inches outside diameter,	360
Diameter of fire box and grate,	6 ft.
Distance of grate to lower tube sheet,	3 ft. 6 in.
Width of air spaces and metal bars in grates,	Air 1-2 in., metal 5-8 in.
Area of water-heating surface,	2,521 sq. ft.
Area of steam-heating surface,	880 sq. ft.
Area of total heating surface,	3,401 sq. ft.
Area of grate surface,	56.7 sq. ft.
Area through tubes,	8 sq. ft.
Ratio of water-heating surface to grate surface,	44.5 to 1
Ratio of steam-heating surface to grate surface,	15.7 to 1
Ratio of total heating surface to grate surface,	60.2 to 1
Ratio of grate surface to tube area,	7.1 to 1

Boiler No. 60 consists of a plant of two vertical boilers, the general features of which are shown in elevation in the following cut. These boilers are of the fire box type and they contain a large area of heating surface, the proportion of which to grate surface is 60.2 to 1. One quarter of this surface is steam heating surface. There is no brick work about the setting, except the brick ash pit which supports the boiler. The outside of the shell is protected by a covering of asbestos board and hair felting. The interior surfaces of the boiler were free from scale.

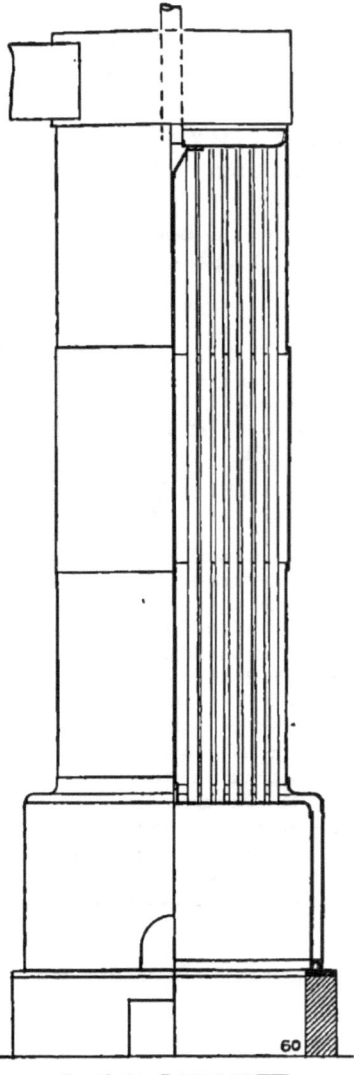

BOILER NO. 60, VERTICAL ELEVATION AND SECTION.

Results of Test, Boiler No. 60.

Test No. 123.

Manner of start and stop,	Thin fire.
Kind of run,	Ordinary.
Duration,	10.3 hrs.
Coal consumed, dry (including wood equivalent),	7,588 lbs.
Percentage of ash,	7.7 per cent.
Water evaporated,	81,730 lbs.
Coal per hour,	739.3 lbs.
Coal per hour per square foot of grate,	13.08 lbs.
Water per hour,	7,962.8 lbs.
Water per hour per square foot of water-heating surface,	3.2 lbs.
Horse-power developed,	243.1 H. P.
Boiler pressure,	46.4 lbs.
Temperature of feed-water,	186 deg.
Temperature of escaping gases,	427 deg.
Number of degrees of superheating,	18 deg.
Draught suction,	0.28 in.
Water per pound of coal,	10.77 lbs.
Water per pound of coal from and at 212 degrees,	11.34 lbs.
Water per pound of combustible from and at 212 degrees,	12.29 lbs.

The test on Boiler No. 60 shows the performance of a well proportioned vertical tubular boiler, using Cumberland bituminous coal. The conditions, both in the matter of proportions of the boiler, and the character of the work done during the test, are favorable to economy. The proportion of water-heating surface to grate surface is as large as good practice in horizontal boilers requires; there is a considerable amount of steam-heating surface; there is a minimum chance for external radiation; there are no avenues for the entrance of air except those provided in the fire-door; and during the test the rate of coal consumption was sufficient to secure excellent combustion.

The result of the test bears out the expectations which these favorable conditions justify. If an allowance be made for the superheated condition of the steam, the equivalent evaporation per pound of combustible from and at 212 degrees becomes 12.40 pounds, and the high character of the economy thus shown needs no comment.

Boiler No. 61.

Kind of boiler,	Cast-iron sectional.
Number in use,	One.
Horse-power (basis 12 square feet),	Thirty-seven.
Kind of coal,	Anthracite broken.
Age,	Ten years.

Boiler No. 61 consists of a number of hollow cast iron sections of spherical form communicating with each other, the lower sections being filled with water and the upper ones furnishing steam space. The products of combustion encircle both the upper and lower sections, and the boiler is thus made a superheating boiler. The general features of the boiler are shown in the following cut.

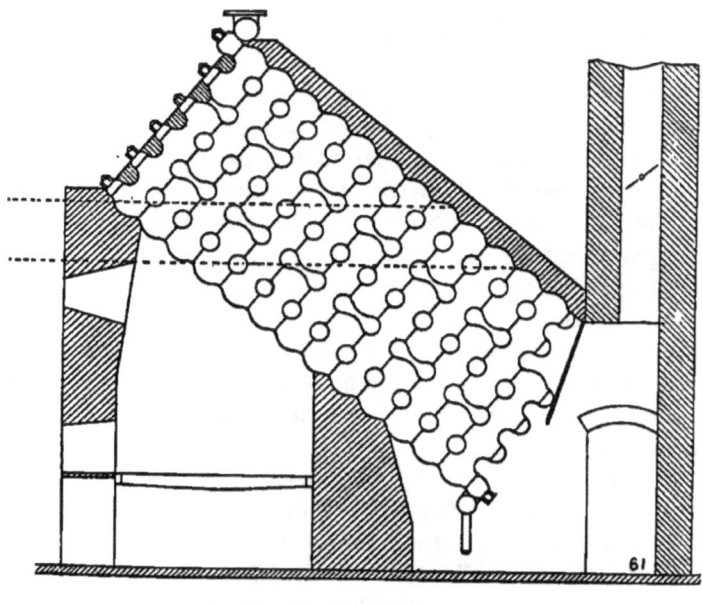

BOILER NO. 61, LONGITUDINAL SECTION.

Dimensions of Boiler No. 61.

Number of sections 5x76,		380
Area of total heating surface,		443 sq. ft.
Area of grate surface,		16.29 sq. ft.
Area through flue,		.97 sq. ft.
Ratio total heating surface to grate surface,		27.2 to 1
Ratio of grate to flue,		16.8 to 1
Chimney height,		100 ft.
	Test No. 124.	Test No. 125.
Area of water-heating surface, square feet,	321	221
Area of steam-heating surface, square feet,	122	222
Ratio of water-heating surface to grate,	19.7 to 1	13.6 to 1

Results of Tests, Boiler No. 61.

	Test No. 124.	Test No. 125.
Duration, ... hrs.	11	11
Coal consumed, ... lbs.	1,600	1,579
Percentage of ash, ... per cent.	9.1	10
Water evaporated, ... lbs.	12,817	12,434
Coal per hour, ... lbs.	145.5	143.5
Coal per hour per square foot of grate, . lbs.	8.9	8.9
Water per hour, ... lbs.	1,165.2	1,130.4
Water per hour per square foot of water-heating surface, ... lbs.	3.6	5.1
Horse-power developed, ... H. P.	37.5	36.6
Boiler pressure, ... lbs.	28.3	28.3
Temperature of feed-water, . . . deg.	123	117
Temperature of escaping flue gases, . deg.	575	540
Number of degrees of superheating, . deg.	25	153
Water per pound of coal, ... lbs.	8.01	7.87
Water per pound of coal from and at 212 degrees, ... lbs.	8.90	8.79
Water per pound of combustible from and at 212 degrees, ... lbs.	9.79	9.78

NOTE. — The tests commenced with a banked fire, and ended 24 hours afterward with the fire in practically the same condition.

The special object of the tests on Boiler No. 61 was to determine its performance as a superheating boiler with two widely different amounts of superheating. The variation in superheating was obtained by carrying the water at two different heights, the two points being indicated in the cut. On the first test the superheating amounted to 25 degrees, while on the second test it amounted to 153 degrees. In spite of the wide difference in the quality of the steam, the evaporation per pound of combustible was the same in both cases.

The ratio of water heating surface to grate surface in this boiler is small, and, as a consequence, the products of combustion escape to the chimney at a high temperature, and the economic result is inferior to that obtained where these conditions are more favorable.

Boiler No. 62.

Kind of boiler,	Cast-iron sectional.
Number used,	Ten.
Horse-power (collective, basis 12 square feet),	Four hundred and sixty.
Kind of coal,	Anthracite Chestnut No. 2.
Age,	Twelve years.

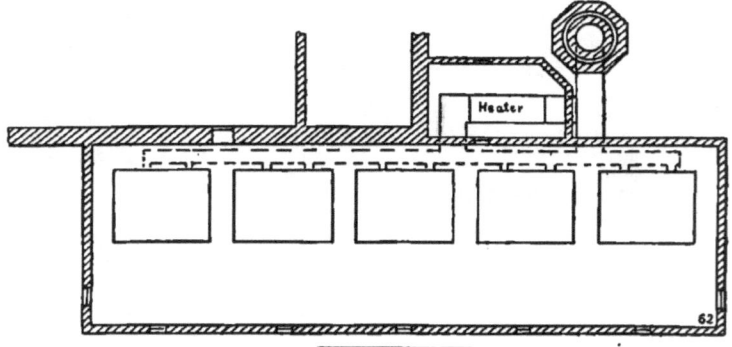

Boiler No. 62, Ground Plan showing Location of Boilers and Flue Heater.

Boiler No. 62 embraces a plant of eight cast-iron sectional boilers, a ground plan of which is shown in the following cut. The individual boilers are of the same general form as that shown in elevation in the cut of Boiler No. 61. In these boilers a line of horizontal plates, or shields, is introduced between the sections, along the water line, in order to separate the space above the water line from that below it, and cut off the steam heating surface. The total area of surface both above and below the water line amounts to 850 square feet in each boiler, and of this, approximately 548 square feet lies below the shields. The upper ends of the boilers are not sufficiently covered with brick work to wholly protect the iron work of the upper sections. The brick walls of the settings are unsound, and many small openings are thus provided for the entrance of air into the furnace space. The plant is fitted with a flue heater, located at the point shown in the cut. The heater consists of a cluster of vertical cast-iron pipes arranged in the same manner as that connected with Boiler No. 33.

Dimensions of Boiler No. 62.

Area of heating surface (below shields),	5,480 sq. ft.
Area of grate surface,	259 sq. ft.
Area through flue,	18 sq. ft.
Area through tubes,	22.5 sq. ft.
Height of chimney,	110 ft.
Width of air spaces and metal bars in grates,	3-8 in.
Ratio of heating surface below shields to grate surface,	21.2 to 1
Ratio of total heating surface to grate surface,	32.8 to 1
Ratio of grate surface to flue area,	11.5 to 1
Area of heating surface in flue heater,	1,280 sq. ft.

Results of Tests, Boiler No. 62.

	Test No. 126.	Test No. 127
Conditions as to flue heater,	Heater in use.	Heater not in use.
Manner of start and stop and kind of run,	Ordinary.	Ordinary.
Duration, hrs.	11	11
Coal consumed, dry (including wood equivalent), lbs.	25,500	28,486
Percentage of ash, per cent.	15.5	15.2
Water evaporated, lbs.	192,862	196,458
Coal per hour, lbs.	2,272.1	2,589.5
Coal per hour per square foot of grate, lbs.	9.2	10.0
Water per hour, lbs.	19,171	20,538
Water per hour per square foot of heating surface, lbs.	2.5	2.6
Horse-power developed, H. P.	568	608.6
Boiler pressure, lbs.	82	82
Temperature of feed-water entering heater, deg.	111	–
Temperature of feed-water entering boiler, deg.	169	112
Temperature of escaping gases leaving boiler, deg.	403	–
Temperature of escaping gases entering chimney, deg.	299	434
Number of degrees flue heater added to water, deg.	58	–
Number of degrees flue heater reduced gases, deg.	104	–
Water per pound of coal, lbs.	7.53	6.89
Water per pound of coal from and at 212 degrees, lbs.	–	7.85
Water per pound of combustible from and at 212 degrees, lbs.	–	9.26

NOTE. — The coal in Test No. 126 when fired contained 5.5 per cent. of moisture; that in No. 127 contained 3.6 per cent.

The tests on Boiler No. 62 had for a principal object the the determination of the economy produced by the use of a flue heater. Test No. 126 was made with the flue heater in use, and Test No. 127 with the heater shut off, and the gases passing to the chimney through the direct flue. The use of this apparatus secured a reduction of 104 degrees in the temperature of the waste gases, or 135 degrees, when referred to the temperature obtained on Test No. 127. The heater increased the temperature of the water 58 degrees, the initial temperature being 111 degrees. The water evaporated per pound of coal was increased by using the heater from 6.89 to 7.53, or 0.64 pounds, and this represents a gain of 9.3 per cent. When the same heater was supplied with water at 200 degrees, the flue gases entering at 380 degrees, the temperature of the water was raised 25 degrees. Looking at the general results of Test No. 127, it appears to be little better than that obtained on the plain cylinder boiler, Test No. 94, in which the temperature of the waste gases was higher to the extent of 133 degrees. It appears that the temperature of the gases, which was 434 degrees on Test No. 127, does not represent the true loss at the chimney, owing to the leakage of a large amount of air into the furnace. As an indication of the extent to which this leakage occurred, it may be said that the draught suction in the main flue, amounting to 0.65 of an inch water pressure, was reduced to 0.34 of an inch in the furnace of one of these boilers, with the fire-doors and ash-pit doors tightly closed and the damper wide open. Had there been no leakage, the full suction of the main flue would, under these circumstances, have been secured at the furnace.

Boiler No. 63.

Kind of boiler,	Cast-iron sectional.
Number used,	Two.
Horse-power (collective, basis 12 square feet),	One hundred and forty.
Kind of coal,	Bituminous Cambria.
Age,	Fifteen years.

Boiler No. 63 embraces a plant of two sectional boilers set in one battery of brick work. The general form of the boilers and the arrangement of the setting are the same as shown in the cut of Boiler No. 61. The whole exterior of the various sections is exposed to the heat, and the surface above the water line is thus made steam heating surface. The boilers had been in use several years, and their interior surfaces had become coated to some extent with scale, which had never been removed.

Dimensions of Boiler No. 63.

Area of water-heating surface,	1,687 sq. ft.
Area of steam-heating surface,	562 sq. ft.
Area of grate surface,	69.7 sq. ft.
Width of air spaces and metal bars in grates.	1-2 in.
Ratio of water-heating surface to grate surface.	24.2 to 1
Ratio of steam-heating surface to grate surface.	8 to 1

Results of Test, Boiler No. 63.

		Test No. 128.
Manner of start and stop and kind of run,		Ordinary.
Duration,	10 4	hrs.
Coal consumed, dry (including wood equivalent),	6.580	lbs.
Percentage of ash,	12.4	per cent.
Water evaporated,	48.059	lbs.
Coal per hour,	660.2	lbs.
Coal per hour per square foot of grate,	9.47	lbs.
Water per hour,	4.641.9	lbs.
Water per hour per square foot of water-heating surface.	2.8	lbs.
Horse-power developed,	155.2	H. P.
Boiler pressure.	54.7	lbs.
Temperature of feed-water,	92.6	deg.
Temperature of escaping gases.	462	deg.
Number of degrees of superheating,	29	deg.
Draught suction,	0.09	in.
Water per pound of coal,	7.30	lbs.
Water per pound of coal from and at 212 degrees,	8.41	lbs.
Water per pound of combustible from and at 212 degrees,	9.61	lbs.

NOTE. — The coal when fired contained 3.7 per cent. of moisture.

The test on Boiler No. 63 shows the performance of a cast-iron sectional boiler, using Cambria bituminous coal. This boiler worked under many unfavorable conditions, and the economic result of the test is inferior to the best practice. The ratio of water heating surface to grate surface is small, and the efficiency of this surface had become reduced by age, so that the escaping gases passed to the chimney at a high temperature. The actual loss due to the waste heat is no doubt greater than would appear from the simple indication of temperature, on account of the facility with which air finds entrance into the furnace in this form of boiler, especially in the case under consideration, where the brick work of the setting had become deteriorated. The steam heating surface produced a superheating amounting to 29 degrees. The boilers developed somewhat more than their rated capacity, with a draught suction of only .09 inch.

Boiler No. 64.

Kind of boiler,	Water-tube.
Number used,	One.
Horse-power (basis 12 square feet),	Seventy.
Kind of coal,	Anthracite Lehigh, chestnut.
Age,	Six months.

Boiler No. 64 is a water tube boiler, the general features and the manner of setting of which are shown in longitudinal section in the following cut. This boiler consists essentially of a cluster of parallel tubes, connected at either end through headers, with a drum above. The water fills the whole up to about the middle of the drum, and the arrangement is such that, in process of operation, a circulation is established through the tubes, the water starting from the back end of the drum, thence passing into the tubes and returning through them to the front end. The products of combustion on leaving the furnace pass around and between the tubes, taking a direction at right angles to their length. Their course lies successively through three compartments, into which the space is divided by means of partitions. The lower half of the drum is exposed to the heat, and the upper half is covered with brick work. The brick side walls of the furnace in this case are provided with perforated plates, through which air is admitted above the fuel, the air first passing back and forth through ducts in the walls.

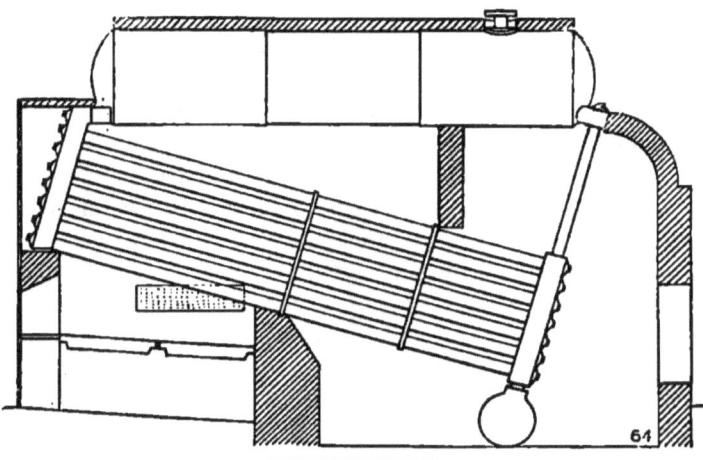

BOILER NO. 64, LONGITUDINAL SECTION.

Dimensions of Boiler No. 64.

Number of tubes 4 inches outside diameter,	42
Length of tubes,	16 ft.
Diameter of drum (one),	30 in.
Mean length of drum,	15 ft., 9 in.
Area of heating surface,	840 sq. ft.
Area of grate surface,	22 5 sq. ft.
Width of air spaces and metal bars in grates,	1-4 in.
Area through flue and chimney,	4 sq. ft.
Height of chimney,	57 ft.
Ratio of heating surface to grate surface,	37.3 to 1
Ratio of grate to flue area,	5.6 to 1

Results of Test, Boiler No. 64.

Test No. 129.

Manner of start and stop and kind of run,		Ordinary.
Duration,	10.7	hrs.
Coal consumed, (including wood equivalent),	2,252	lbs.
Percentage of ash,	14	per cent.
Water evaporated,	19,178	lbs.
Coal per hour,	209.5	lbs.
Coal per hour per square foot of grate,	9.3	lbs.
Water per hour,	1,784	lbs.
Water per hour per square foot of heating surface,	2.1	lbs.
Horse-power developed,	54.8	H. P.
Boiler pressure,	89	lbs.
Temperature of feed-water,	180	deg.
Temperature of escaping gases,	337	deg.

Draught suction, ... 0.09 in.
Water per pound of coal, ... 8.52 lbs.
Water per pound of coal from and at 212 degrees, . . 9.12 lbs.
Water per pound of combustible from and at 212 degrees, 10.61 lbs.

The test on Boiler No. 64 shows the performance of a water tube boiler, working under somewhat unfavorable conditions for the best results. The slow rate of combustion employed, and the admission of air above the fuel, the fuel being in this case anthracite coal, were no doubt the cause of some loss. As the results stand they do not differ materially in point of economy from those obtained from the horizontal tubular boilers, No. 13, No. 14, and No. 15, made with a similar class of coal.

Boiler No. 65.

Kind of boiler, ... Water-tube.
Number used, ... Four.
Horse-power (collective, basis 12 sq. ft.), Four hundred and sixty-eight.
Kind of coal, ... Anthracite Shamokin, pea.
Age, ... Three years.

Boiler No. 65 embraces four water tube boilers of the general form shown in longitudinal section in the following cut. They are set in two independent batteries of brick work, each embracing two boilers. The main flue behind the boilers is provided with a small flue heater, consisting of wrought iron pipes, having an exposed surface amounting to 515 square feet.

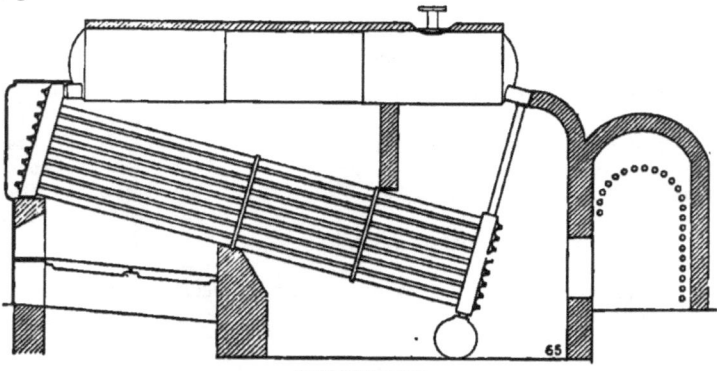

BOILER No. 65, LONGITUDINAL SECTION.

Dimensions of Boiler No. 65.

Number of tubes (collective) 4 inches outside diameter,	256
Length of tubes,	18 ft.
Diameter of drums (four),	36 in.
Mean length of drums,	17 ft. 6 in.
Area of heating surface,	5,614 sq. ft.
Area of grate surface,	141.7 sq. ft.
Area through flue,	18.5 sq. ft.
Width of air spaces and metal bars in grates,	Air 5-16 in., metal 7-16 in.
Ratio of heating surface to grate surface,	40 to 1
Ratio of grate surface to flue area,	7.2 to 1

Results of Test, Boiler No. 65.

Test No. 130.

Manner of start and stop and kind of run,	Ordinary.
Duration,	11 hrs.
Coal consumed, dry (including wood equivalent),	19,043 lbs.
Percentage of ash,	17.4 per cent.
Water evaporated,	161,656 lbs.
Coal per hour,	1,731.2 lbs.
Coal per hour per square foot of grate,	12.2 lbs.
Water per hour,	14,696 lbs.
Water per hour per square foot of heating surface,	2.5 lbs.
Horse-power developed,	474.4 H. P.
Boiler pressure,	101.2 lbs.
Temperature of feed-water leaving heater,	145.3 deg.
Temperature of escaping gases,	353 deg.
Draught suction,	0.29 in.
Percentage of moisture in steam,	0.6 per cent.
Water per pound of coal,	8.49 lbs.
Water per pound of coal from and at 212 degrees,	9.45 lbs.
Water per pound of combustible from and at 212 degrees,	11.44 lbs.

NOTE.—The coal when fired contained 5 per cent. of moisture.

The test on Boiler No. 65 shows the performance of a water tube boiler with one of the small grades of anthracite coal. The conditions with respect to capacity, rate of combustion and temperature of escaping gases, are all favorable, and an exceedingly high evaporative result, based on combustible, is secured. A subsequent test was made when only three boilers were in use, developing about the same total power. The rate of combustion was 15.7 pounds, the percentage of ash 15.8 per cent., the temperature of the feed water 156.7 degrees, and the temperature of the escaping gases 389 degrees. The evaporation was 8.48 pounds of water per pound of coal, and 11.08 pounds of water from and at 212 degrees per pound of combustible.

The temperature of the escaping gases was obtained at a point between the flue heater and the chimney. The heater added 18 degrees to the temperature of the water on Test No. 130, that on entering the heater being 126.8 degrees.

Boiler No. 66.

Kind of boiler,	Water tube.
Number used,	One.
Horse power (basis 12 square feet),	One hundred.
Kind of coal,	Anthracite Lehigh, broken.
Age,	Six months.

Boiler No. 66 is a water tube boiler, having the general features shown in longitudinal section in the following cut.

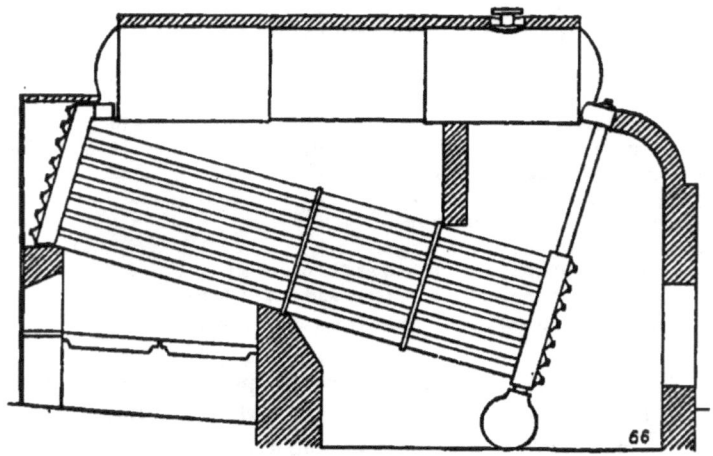

BOILER No. 66, LONGITUDINAL SECTION.

Dimensions of Boiler No. 66.

Number of tubes,	54	
Length of tubes,	18	ft.
Diameter of drum,	36	in.
Mean length of drum,	17 ft. 6	in.
Area of heating surface,	1,196	sq. ft.
Area of grate surface,	28.7	sq. ft.

Area through flue, ... 9.8 sq. ft.
Width of air spaces and metal bars in grates, . Air 3-8 in., metal 1-4 in.
Ratio of heating surface to grate surface, ... 40.3 to 1
Ratio of grate to flue area, ... 6.1 to 1

Results of Test, Boiler No. 66.

	Test No. 131.
Manner of start and stop and kind of run,	Ordinary.
Duration,	10.5 hrs.
Coal consumed,	5,553 lbs.
Percentage of ash,	9.2 per cent.
Water evaporated,	45,803 lbs.
Coal per hour,	528.8 lbs.
Coal per hour per square foot of grate,	17.8 lbs.
Water per hour,	4,362.2 lbs.
Water per hour per square foot of heating surface,	3.6 lbs.
Horse-power developed,	133.1 H. P.
Boiler pressure,	.73 lbs.
Temperature of feed water,	183 deg
Temperature of escaping gases,	540 deg.
Water per pound of coal,	8.24 lbs.
Water per pound of coal from and at 212 degrees,	8.77 lbs.
Water per pound of combustible from and at 212 degrees,	9.68 lbs.

The test on Boiler No. 66 shows the performance of a water tube boiler, which had favorable proportions and used a standard grade of coal, but which was operated under unfavorable conditions in respect to capacity. The rate of combustion was 17.8 pounds of coal per square foot of grate per hour, which, for the class of fuel used, is too high for economical work. The horse power developed was one-third above the rating. The evaporative result is low compared with the best practice, and the high temperature of the escaping gases, which was 540 degrees, reveals the cause of the inferior performance. At the close of the test it was found that some of the partitions were disarranged, thereby allowing at certain points a free passage of the products of combustion along the tubes to the flue. This defect was largely the cause of the high temperature of the gases.

Boiler No. 67.

Kind of boiler,	Water tube.
Number used,	One.
Horse-power (basis 12 square feet),	Seventy.
Kind of coal,	Anthracite Lehigh, chestnut No. 2.
Age,	Six months.

Boiler No. 67 is a water tube boiler, having the general features shown in the cut of Boiler No. 66. It is of comparatively small size, having only 42 tubes, and it is connected with a chimney which is deficient in draught power.

Dimensions of Boiler No. 67.

Number of tubes 4 inches outside diameter,	42
Length of tubes,	16 ft.
Area of heating surface,	839 sq. ft.
Area of grate,	23 sq. ft.
Area through flue,	5.3 sq. ft.
Width of air spaces and metal bars in grates,	Air 5-16 in., metal 1-2 in.
Height of chimney,	60 ft.
Ratio of heating surface to grate surface,	36.5 to 1
Ratio of grate surface to flue area,	4.3 to 1

Results of Test, Boiler No. 67.

	Test No. 132.
Manner of start and stop,	Ordinary.
Kind of run,	Continuous.
Duration,	8.3 hrs.
Coal consumed, dry (including wood equivalent),	1,576 lbs.
Percentage of ash,	14.7 per cent.
Water evaporated,	11,292 lbs.
Coal per hour,	189.2 lbs.
Coal per hour per square foot of grate,	8.2 lbs.
Water per hour,	1,355.5 lbs.
Water per hour per square foot of heating surface,	1.6 lbs.
Horse-power developed,	47.2 H. P.
Boiler pressure,	104 lbs.
Temperature of feed-water,	68 deg.
Temperature of escaping gases,	360 deg.
Draught suction,	0.2 in.
Water per pound of coal,	7.16 lbs.
Water per pound of coal from and at 212 degrees,	8.52 lbs.
Water per pound of combustible from and at 212 degrees,	10.00 lbs.

The test on Boiler No. 67 shows the performance of a small water tube boiler, using one of the low grades of anthracite coal. The result obtained, which was 10 pounds of water from and at 212. per pound of combustible, is somewhat inferior to good practice. Test No. 47, made with a similar coal on a horizontal tubular boiler, gave 10.72 pounds and Test No. 58, 10.78 pounds. In the case under consideration the boiler had plenty of heating surface, and the heat was well absorbed, as the comparatively low flue temperature indicates. The rate of combustion is somewhat low for the water tube type of boiler, the draught being deficient, and this, coupled with the possibility of an inferior quality of fuel used furnishes a partial, and perhaps a full explanation of the inferior character of the result.

Boiler No. 68.

Kind of boiler,	Water tube.
Number used,	Two.
Horse-power (collective, basis 12 sq. ft.),	Two hundred and sixty.
Kind of coal,	George's Creek Cumberland.
Age,	Two years.

Boiler No. 68 consists of a plant of two water tube boilers of the form shown in longitudinal section in the cut of Boiler No. 64. The brick side walls of the furnaces are fitted with perforated tiles, through which air is admitted over the fuel, as shown in the cut referred to. This plant is provided with a flue heater, the location of which with reference to the boilers and chimney is shown in the following cut. The heater consists of a cluster of vertical cast iron pipes, similar to that in use on Boiler No. 33. The heating surface exposed by these pipes is about half as much as the total heating surface of the two boilers.

Dimensions of Boiler No. 68.

Number of tubes 4 inches outside diameter,	144
Length of tubes,	18 ft.
Diameter of drums (2),	36 in.
Mean length of drums,	17 ft., 6 in.
Area of heating surface,	3,126 sq. ft.
Area of grate surface,	50 sq. ft.
Area through flue,	11 sq. ft.
Height of chimney,	80 ft.
Width of air spaces and metal bars in grates,	Air 1-2 in., metal 11-16 in.
Ratio of heating surface to grate surface,	62.5 to 1
Ratio of grate surface to flue area,	4.5 to 1
Area of heating surface in flue heater,	1,600 sq. ft.

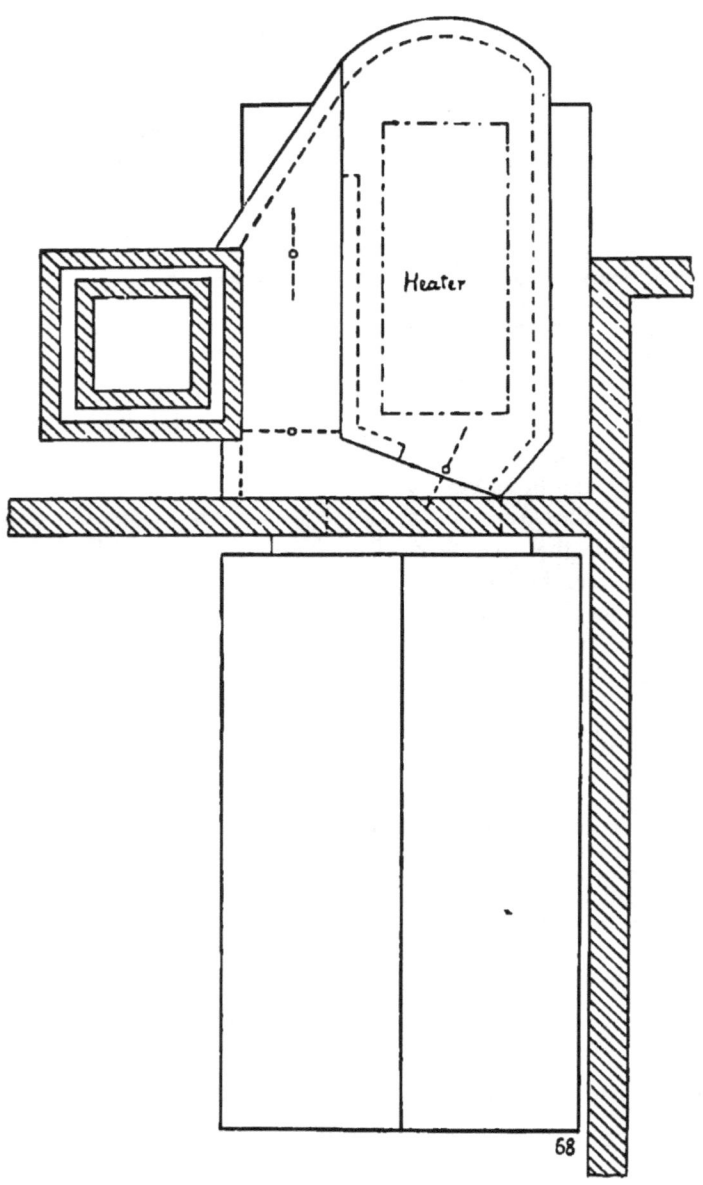

Boiler No. 68, Ground Plan showing Location of Boilers and Flue Heater.

Results of Tests, Boiler No. 68.

	Test No. 133.	Test No. 134.
	Heater in use.	Heater not in use.
Manner of start and stop and kind of run.	Ordinary.	Ordinary.
Duration, ... hrs.	11.5	11.5
Coal consumed, dry (including wood equivalent), ... lbs.	8,743	9,694
Percentage of ash, . . . per cent.	7.5	7.7
Water evaporated, ... lbs.	84,078	82,725
Coal per hour, ... lbs.	760.3	843
Coal per hour per square foot of grate, . lbs.	15.2	16.8
Water per hour, ... lbs.	7,310.4	7,193.2
Water per hour per square foot of heating surface, ... lbs.	2.3	2.3
Horse-power developed, . . . H. P.	247	243.5
Boiler pressure . . . . . lbs.	68	67
Temperature of feed-water entering heater, deg.	84	–
Temperature of feed water entering boiler, deg.	196	82
Temperature of escaping gases leaving boiler, deg.	435	452
Temperature of escaping gases leaving heater, deg.	279	–
Draught suction, ... in.	0.25	0.27
Percentage of moisture in steam, per cent.	1.3	–
Water per pound of coal, ... lbs.	9.62	8.53
Water per pound of coal from and at 212 degrees, ... lbs.		9.95
Water per pound of combustible from and at 212 degrees, ... lbs.		10.79

NOTE. — The coal when fired contained 5 per cent. of moisture.

The tests on Boiler No. 68 had for a principal object the determination of the economy produced by the use of a flue heater. The heater was in operation on Test No. 133, and it was shut off on Test No. 134, the flue gases passing directly from the boiler to the chimney. The use of this apparatus secured an increase in the temperature of the water of 196° — 84° = 112 degrees, and a reduction in the temperature of the gases of 435° — 279° = 156 degrees, resulting in an increase in the evaporation per pound of coal amounting to 12.8 per cent.

Looking at the general performance of the boilers, when worked without the heater, it appears that the result secured is much inferior in point of economy to that obtained from the best class of shell boilers. The proportion of heating surface to grate surface is ample, so also is the rate of combustion. The temperature of the escaping gases is not excessive. The quality of the coal, judging from the low percentage of ash, is excellent. With all these conditions, which would ordinarily be considered favorable, it is difficult to assign a cause for the inferior result, unless it be the loss produced by the admission of air over the fuel through the passages in the walls, and that which may have entered through openings in the brick work due to the somewhat long service to which the boiler had been subjected. The admission of air in this manner makes the actual loss from the waste heat in the gases greater than that apparently indicated by the temperature.

Previous to the introduction of the flue heater a test was made upon the plant when the boilers had been in use but a few weeks, the coal employed being Powelton bituminous mixed with one-fourth of its weight of pea and dust coal. The grate surface on this occasion had a much larger area, being 70 square feet and the proportion of heating surface to grate surface was thereby reduced to 44.7 to 1. The principal results were as follows:

Coal per hour per square foot of grate surface,	16.7	lbs.
Percentage of ash,	9.0	per cent.
Water per hour per square foot of heating surface,	3.7	lbs.
Horse-power developed,	403.3	H. P.
Boiler pressure,	77.0	lbs.
Temperature of feed-water,	38.0	deg.
Temperature of escaping gases,	402.0	deg.
Draught suction,	0.5	in.
Water per pound of coal,	9.75	lbs
Water per pound of coal from and at 212 degrees,	11.86	lbs
Water per pound of combustible from and at 212 degrees,	13.01	lbs

NOTE.— The coal when fired contained 4.5 per cent. of moisture.

This test is remarkable on account of the high character of the result. Compared with the later test it shows the effect which age may produce upon the performance of a boiler. Although the boiler developed nearly 50 per cent. more than that on test No. 134, the heat was so much better absorbed by the new and clean surfaces that the temperature of the gases was reduced to 402 degrees, which, under the circumstances, is most favorable to economy. Futhermore, the new condition of the setting prevented the introduction of unneeded air, which seems to have been one of the main causes of the unfavorable results of the later tests.

Boiler No. 69.

Kind of boiler,	Water tube.
Number used,	One.
Horse-power (basis 12 square feet),	Eighty.
Kind of coal,	Anthracite Lackawanna Chestnut No. 2.
Age,	Three months.

Boiler No. 69 is a water tube boiler, having the general features shown in longitudinal section in the following cut. It is provided at the top with three drums, 14 inches in diameter, connected at the ends to the tubes below, and these are wholly enclosed in the chamber formed by the brick setting. The water line is carried to about the central point of the drums, and the exterior surface above it furnishes a small amount of steam heating surface.

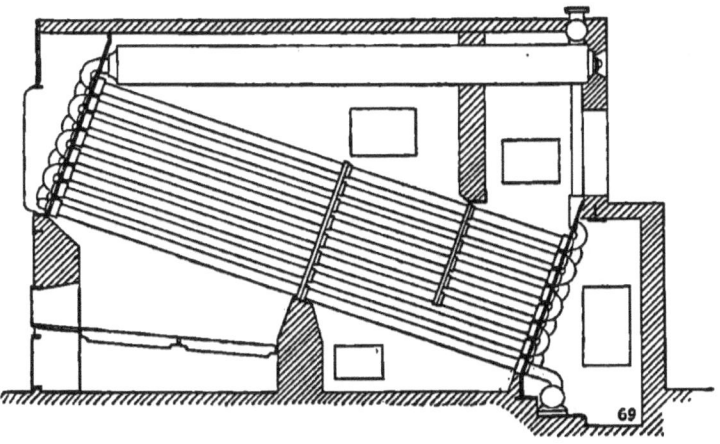

BOILER NO. 69, LONGITUDINAL SECTION.

BOILER No. 69.

Dimensions of Boiler No. 69.

Number of tubes 4 inches outside diameter,	48
Length of tubes,	15 ft.
Diameter of drums (three),	14 in.
Length of drums,	15 ft.
Area of water-heating surface,	848 sq. ft.
Area of steam-heating surface,	114 sq. ft.
Area of total-heating surface,	962 sq. ft.
Area of grate surface,	27 sq. ft.
Area through flue,	3.7 sq. ft.
Height of chimney,	70 ft.
Width of air spaces and metal bars in grates, Air 3-8 in., metal 1-2 in.	
Ratio of water-heating surface to grate surface,	31.4 to 1
Ratio of steam-heating surface to grate surface,	4.2 to 1
Ratio of grate surface to flue area,	7.4 to 1.

Results of Test, Boiler No. 69.

Test No. 135.

Manner of start and stop and kind of run,	Ordinary.
Duration,	10.1 hrs.
Coal consumed, (including 91 pounds leather scraps taken to be equal to 91 pounds of coal),	2,972 lbs.
Percentage of ash,	16.4 per cent.
Water evaporated,	23,396 lbs.
Coal per hour,	294 lbs.
Coal per hour per square foot of grate,	10.9 lbs.
Water per hour,	2,314.1 lbs.
Water per hour per square foot of water-heating surface,	2.7 lbs.
Horse-power developed,	73.9 H. P.
Boiler pressure,	84 lbs.
Temperature of feed-water,	150 deg.
Temperature of escaping gases,	428 deg.
Draught suction,	0.12 in.
Water per pound of coal,	7.87 lbs.
Water per pound of coal from and at 212 degrees,	8.66 lbs.
Water per pound of combustible from and at 212 degrees,	10.36 lbs.

NOTE.—When the water was carried below the ordinary level, the steam was superheated a few degrees. Ordinarily the thermometer failed to show any superheat.

The test on Boiler No. 69 shows the performance of a water tube boiler, using one of the small grades of anthracite coal. The ratio of heating surface to grate surface is somewhat small for a water tube boiler. As a result, the temperature of the waste gases is above the point of the best efficiency, and the evaporative result, which is 10.36 pounds of water from and at 212 per pound of combustible, is somewhat unfavorable.

Little effect seems to have been produced by the steam heating surface, so far as it could be observed from the thermometer immersed in the steam.

Boiler No. 70.

Kind of boiler,	Water tube.
Number used,	One.
Horse-power (basis 12 sq. ft.),	One hundred and thirty-six.
Kind of coal,	Bituminous Cambria.
Age,	Eight years.

Boiler No. 70 is a water tube boiler having the general features shown in the cut of Boiler No. 66.

Dimensions of Boiler No. 70.

Number of tubes 4 inches outside diameter,	80
Length of tubes,	18 ft.
Diameter of drums (two),	30 in.
Mean length of drums,	7 ft. 6 in.
Area of heating surface,	1,638 sq. ft.
Area of grate surface,	36 sq. ft.
Width of air spaces and metal bars in grates,	3-4 in.
Ratio of heating surface to grate surface,	45.5 to 1

Results of Tests. Boiler No. 70.

Test No. 136

Manner of start and stop and kind of run,	Ordinary.
Duration,	10.6 hrs.
Coal consumed, dry (including wood equivalent),	6,185 lbs.
Percentage of ash,	10.5 per cent.
Water evaporated,	52,206 lbs.
Coal per hour,	608.9 lbs.
Coal per hour per square foot of grate,	16.89 lbs.
Water per hour,	4,946 lbs.
Water per hour per square foot of heating surface,	3 lbs.
Horse-power developed,	166.3 H. P.
Boiler pressure,	62.4 lbs.
Temperature of feed-water,	88.3 deg.
Temperature of escaping gases,	471 deg.
Draught suction,	0.2 in.
Percentage of moisture in steam,	0.42 per cent.
Water per pound of coal,	8.44 lbs.
Water per pound of coal from and at 212 degrees,	9.78 lbs.
Water per pound of combustible from and at 212 degrees,	10.93 lbs.

NOTE. — The coal when fired contained 3.7 per cent. of moisture.

The test on Boiler No. 70 shows the performance of a water tube boiler, using Cambria bituminous coal. The power developed is somewhat more than its nominal capacity, and a high rate of combustion prevails. The temperature of the escaping gases is above the economical limit, and the evaporative result is correspondingly low. The quantity of ash contained in the coal, viz. 10.5 per cent., is higher than is found in the best grades of bituminous coal, and to this indication of inferior quality of fuel may be attributed in some measure the low degree of economy which was obtained.

Boiler No. 71.

Kind of boiler,	Water tube.
Number used,	Two.
Horse-power (collective, basis 12 sq. ft.),	One hundred and fifty-seven.
Kind of coal,	George's Creek Cumberland.
Age,	Four years.

Boiler No. 71 is a water tube boiler, having the general features shown in the cut of Boiler No. 66. It embraces two boilers set in one battery of brick work.

Dimensions of Boiler No. 71.

Number of tubes 4 inches outside diameter,	96	
Length of tubes,	16	ft.
Diameter of drums (two),	30	in.
Length of drums,	16 ft. 6	in.
Area of heating surface,	1,886	sq. ft.
Area of grate surface,	38.9	sq. ft.
Area through flue,	9.2	sq. ft.
Width of air spaces and metal bars in grates,	Air 5-8 in., metal 3-8	in.
Ratio of heating surface to grate surface,	48.4	to 1
Ratio of grate surface to flue area,	4.3	to 1

Results of Test, Boiler No. 71.

Test No. 137

Manner of start and stop and kind of run,	Ordinary.	
Duration,	10	hrs.
Coal consumed, dry (including wood equivalent),	6,812	lbs.
Percentage of ash,	6.4	per cent.
Water evaporated,	59,113	lbs.
Coal per hour,	634.9	lbs.
Coal per hour per square foot of grate,	16.3	lbs.
Water per hour,	5,510.6	lbs.
Water per hour per square foot of heating surface,	2.9	lbs.

Horse-power developed,	189.3	H. P.
Boiler pressure,	72.9	lbs.
Temperature of feed-water,	66	deg.
Temperature of escaping gases,	523	deg.
Draught suction,	0.21	in.
Percentage of moisture in steam,	0.4	per cent.
Water per pound of coal,	8.68	lbs.
Water per pound of coal from and at 212 degrees,	10.28	lbs.
Water per pound of combustible from and at 212 degrees,	10.98	lbs.

The test on Boiler No. 71 shows the performance of a water tube boiler using Cumberland bituminous coal. The conditions under which the tests were made are all favorable to good economy with the exception of one, that is, the temperature of the escaping gases. This is so far above the economic point that it fully accounts for the somewhat low evaporative result secured, this being 10.98 pounds of water from and at 212 per pound of combustible. It is difficult to assign a satisfactory cause for the high temperature of the waste gases in this case. It is evidently due to inefficiency of heating surface, but the difficulty lies in determining the cause of the inefficiency. It is known that the interior surfaces of the boiler were clean; the exterior surfaces were as clean as the thorough use of a steam jet could make them. Either this method of removing the deposits of soot from the exterior of the pipes did not accomplish its purpose, or there is some fault in the arrangement of the heating surface, whereby it fails to properly absorb the heat.

Preparatory to the test on this boiler the cracks in the brick work, and all the crevices around the doors, with which the boiler is provided for purposes of cleaning, were filled with fire clay. The effect of reducing by this means the quantity of air which was drawn in was revealed by means of the draught gauge. With no fire in the furnace, and the fire and ash doors closed, a draught of 0.31 of an inch in the main flue, with wide open damper, produced a draught of 0.25 of an inch between the damper and the boiler, before the openings were closed, and the full draught of 0.31 of an inch after they were closed. When the damper was in a partly closed position, the draught in the main flue remaining the same, the amount realized was 0.09 of an inch before stopping up the openings, and 0.20 of an inch after stopping them up. The crevices were not fully closed, else the full draught would have been realized when the damper was partially open. Repeated tests, made beforehand, with open crevices, showed a lower economic result than that given here, and this proves that the leakage of too much air into a furnace produces an unfavorable effect upon the economy.

Summary of Tests.

Number of boiler.	Kind of boiler.	Ratio of water-heating surface to grate surface.	Ratio of steam-heating surface to grate surface.	Kind of coal.	Percentage of ash.	Coal per hour per square foot of grate.	Percentage of moisture or degrees of superheating.	Temperature of escaping gases.	Water per pound of combustible from and at 212 degrees.
						Lbs.		Deg. Fahr.	Lbs.
1	Hor. Ret. Tub. with superheater.	33.7 to 1	—	Anthracite, Hazelton, Chestnut.	15.6	7.7	66 deg.	360	10.25
	Hor. Direct Tub.	27.7 to 1	—	Ditto.	—	8.5	None.	367	10.85
2	Hor. Ret. Tub. with pipe surface in furnace.	30.5 to 1	—	Bituminous, Cumberland.	9.6	10.8	0.6 per ct.	355	10.19
	Ditto.	" "	—	Ditto.	10.5	10.0	3 per ct.	346	10.91
3	Hor. Ret. Tub.	93.8 to 1	4.5 to 1	Anthracite Lackawanna, broken.	16.7	11.11	—	482	10.73
4	Hor. Ret. Tub. with double passage of gases.	41.4 to 1	—	Mixture of Anthracite screenings and N. S. Culm.	15.4	7.15	—	305	9.66
5	Hor. Ret. Tub.	32.2 to 1	—	Geo.'s Creek, Cumberland, Bituminous.	11.1	10.1	—	435	11.52
	" "	" "	—	Delaware and Hudson, Lackawana, Anthracite, broken.	14.2	10.1	—	443	10.76
	" "	" "	—	1 part Cumberland, 2 parts Pea and Dust.	21.3	8.3	—	430	10.18
	" "	" "	—	1 part Nova Scotia Culm, 2 parts Pea and Dust.	26.2	9.2	—	406	9.87

Summary of Tests.

Number of boiler.	Kind of boiler.	Ratio of water-heating surface to grate surface.	Ratio of steam-heating surface to grate surface.	Kind of coal.	Percentage of ash.	Coal per hour per square foot of grate.	Percentage of moisture or degrees of superheating.	Temperature of escaping gases.	Water per pound of combustible from and at 212 degrees.
						Lbs.		Deg. Fahr.	Lbs.
6	Hor. Ret. Tub.	33.7 to 1	—	2 parts screenings, 1 part Nova Scotia Culm.	14.3	11.02	—	453	8.69
	" "	" "	—	Ditto.	16.1	11.05	—	461	8.69
7	Hor. Ret. Tub.	37.6 to 1	—	Anthracite screenings 3 parts, Cumberland bituminous 1 part.	14.2	10.1	—	410	10.70
8	Hor. Direct Tub. detached furnace.	33 to 1	—	Anthracite pea and dust. 3 parts, George's Creek Cumberland 1 part.	11	10.5	—	395	10.73
	Ditto.	" "	—	Geo.'s Creek Cumberland, Bituminous.	9.6	12.7	—	400	10.79
9	Hor. Ret. Tub. with water leg front.	34.6 to 1	—	Lehigh Egg, Anthracite.	6.6	18.2	0.3	453	11.17
	Ditto.	" "	—	Equal parts Anthracite Screenings, and Cumberland.	9.4	14	—	349	11.24
	Ditto.	" "	—	Cumberland.	13.5	13.7	—	343	10.85
					7.0	18.8	—	486	10.34

Summary of Tests.

Number of boiler.	Kind of boiler.	Ratio of water-heating surface to grate surface.	Ratio of steam-heating surface to grate surface.	Kind of coal.	Percentage of ash.	Coal per hour per square foot of grate.	Percentage of moisture or degrees of superheating.	Temperature of escaping gases.	Water per pound of combustible from and at 212 degrees.
						Lbs.		Deg. Fahr.	Lbs.
10	Hor. Ret. Tub.	33.5 to 1	4.1 to 1	Anthracite Lehigh, broken.	13.1	4.7	—	297	11.01
	" " "	" "	" "	Ditto.	14.1	4.7	—	299	11.42
	" " "	" "	" "	Ditto.	12.5	8.7	—	340	11.63
	" " "	" "	" "	Ditto.	13.9	8.6	—	348	11.44
11	Hor. Ret. Tub.	35.5 to 1	—	Anthracite broken.	15	5.7	—	—	10.48
	" " "	" "	—	Petroleum oil.	—	—	—	—	—
12	Hor. Ret. Tub. with double passage of gases.	42 to 1	—	Lackawanna broken.	12	11	—	306	11.13
	Ditto.	" "	—	Lehigh broken.	10.1	12.2	—	346	11.20
	Ditto.	" "	—	Geo.'s Creek Cumberland.	6.6	14	—	381	11.37
	Ditto.	" "	—	2 parts Screenings, 1 part George's Creek Cumberland.	16.5	12.2	—	343	11.40
13	Hor. Ret. Tub.	33.7 to 1	—	Anthracite, Lehigh, Chestnut.	16.4	12.1	—	444	10.76
	" " "	" "	—	Ditto.	17.5	6.5	—	356	11.01
14	Hor. Dir. Tub.	27.7 to 1	—	Ditto.	15.7	11.4	—	476	10.46
	" " "	" "	—	Ditto.	15.5	6	—	380	10.88

Summary of Tests.

Number of boiler.	Kind of boiler.	Ratio of water-heating surface to grate surface.	Ratio of steam-heating surface to grate surface.	Kind of coal.	Percentage of ash.	Coal per hour per square foot of grate.	Percentage of moisture or degrees of superheating.	Temperature of escaping gases.	Water per pound of combustible from and at 212 degrees.
						Lbs.		Deg. Fahr.	Lbs.
15	Hor. Ret. Tub. with water leg front.	37 to 1	—	Anthracite, Lehigh Chestnut.	14.4	14.2	—	365	10.47
	Ditto.	" "	—	Ditto.	14.2	10	—	350	10.75
	Ditto.	" "	—	Honeybrook Lehigh Chestnut.	14.2	10	—	350	10.75
	Ditto.	" "	—	Honeybrook Lehigh Pea.	15.8	10.8	—	348	9.98
	Ditto.	" "	—	Honeybrook Lehigh broken.	10.5	13.2	—	372	10.74
	Ditto.	" "	—	Geo.'s Creek Cumberland.	6.6	9.1	—	387	11.78
	Ditto.	" "	—	Pea and Dust 2 parts, Cumberland 1 part.	11.4	10.7	—	387	11.06
	Ditto.	" "	—	Cumberland, Geo.'s Creek.	6.6	9.1	—	387	11.78
	Ditto.	" "	—	Ditto.	6.4	9.4	—	372	11.09
	Ditto.	" "	—	Pea and Dust 2 parts, Cumberland 1 part.	11.4	10.7	—	387	11.06
	Ditto.	" "	—	Ditto.	13.8	10.8	—	362	11.60
	Ditto.	" "	—	Anthracite Lehigh broken.	10	12.2	—	389	10.65
	Ditto.	" "	—	Ditto.	10.5	13.2	—	372	10.74

Summary of Tests.

Number of boiler.	Kind of boiler.	Ratio of water-heating surface to grate surface.	Ratio of steam-heating surface to grate surface.	Kind of coal.	Percentage of ash.	Coal per hour per square foot of grate.	Percentage of moisture or degrees of superheating.	Temperature of escaping gases.	Water per pound of combustible from and at 212 degrees.
16	Hor. Ret. Tub.	33.8 to 1	—	Anthracite stove.	14.7	Lbs. 9.3	—	Deg. Fahr. 314	Lbs. 11.06
	" " "	" "	—	Anthracite Chestnut No. 2.	12.8	9.6	—	312	10.72
	" " "	" "	—	2 parts Pea and Dust, 1 part Clearfield.	14	9.7	—	326	10.85
17	" " "	26.5 to 1	—	Anthracite White Ash, broken.	10.1	12.9	—	455	9.75
	" " "	" "	—	Anthracite Pea.	20.9	11.7	—	448	10.63
	" " "	" "		44 parts Pea and Dust, 37 parts Nova Scotia Culm.	17.9	13.6	—	460	9.34
18	Hor. Ret. Tub. with double furnace.	54.4 to 1	—	Bituminous Cumberland.	8.5	19.3	—	472	10.93
	Ditto.	" "	—	Anthracite Lehigh, broken.	13.5	12.5	—	354	10.05
19	Hor. Ret. Tub.	29.4 to 1	—	Bituminous Cumberland.	8.7	10.9	—	530	10.60
20	Hor. Ret. Tub.	37.1 to 1	—	3 parts Pea and Dust, 1 part Cumberland.	14.1	10.46	—	467	10.23
21	Hor. Ret. Tub.	37.1 to 1	—	Ditto.	14.5	11.46	—	474	10.74

Summary of Tests.

Number of boiler.	Kind of boiler.	Ratio of water-heating surface to grate surface.	Ratio of steam-heating surface to grate surface.	Kind of coal.	Percentage of ash.	Coal per hour per square foot of grate.	Percentage of moisture or degrees of superheating.	Temperature of escaping gases.	Water per pound of combustible from and at 212 degrees.
						Lbs.		Deg. Fahr.	Lbs.
22	Hor. Ret. Tub.	68 to 1	—	Anthracite Lehigh, Egg.	12.9	11.5	—	350	11.03
23	Hor. Ret. Tub.	60 to 1	—	Anthracite Chestnut No. 2.	15.7	10.3	—	316	10.78
24	Hor. Ret. Tub.	26.6 to 1	—	Anthracite broken.	12	7.9	—	474	9.87
25	Hor. Ret. Tub.	44.2 to 1	—	Anthracite White-Ash broken.	15.7	9.4	—	369	10.61
26	Hor. Ret. Tub.	39.8 to 1	—	Nova Scotia Culm.	9.4	14.4	—	528	9.51
27	Hor. Ret. Tub.	39.8 to 1	—	" "	9.7	14.1	—	513	9.65
28	Hor. Ret. Tub.	42.8 to 1	—	" "	8.5	12.6	—	489	9.79
29	Hor. Ret. Tub.	36.8 to 1	—	" "	9.6	12.5	—	480	9.49
30	Hor. Ret. Tub. with smoke burner.	47 to 1	—	Cumberland bituminous.	6.5	13.2	—	362	9.91
31	Hor. Ret. Tub.	41.6 to 1	—	Geo.'s Creek Cumberland.	6.6	7	—	431	12.07
	" "	" "	—	Coke from gas coal.	4.9	8.1	—	428	10.11
32	Hor. Ret. Tub.	40 to 1	—	Geo.'s Creek Cumberland.	6.5	11.1	2.2 per ct.	408	11.98
33	Hor. Ret. Tub. with flue heater.	37.9 to 1	—	2 parts Anthracite Lehigh Buck wheat, 1 part Clearfield Bituminous.	11.8	9.8	—	231	—
	Ditto, without flue-heater.	" "	—	Ditto.	11.8	10.9	—	399	10.45

Summary of Tests.

Number of boiler.	Kind of boiler.	Ratio of water-heating surface to grate surface.	Ratio of steam-heating surface to grate surface.	Kind of coal.	Percentage of ash.	Coal per hour per square foot of grate.	Percentage of moisture or degrees of superheating.	Temperature of escaping gases.	Water per pound of combustible from and at 212 degrees.
						Lbs.		Deg. Fahr.	Lbs.
34	Hor. Ret. Tub.	32.2 to 1	—	Bituminous Walston.	7.3	12.3	—	572	10.00
35	Hor. Ret. Tub.	47.4 to 1	—	Geo.'s Creek Cumberland.	8.3	6.7	0.86 p. c.	340	11.24
36	Hor. Ret. Tub. detached furnace.	" "	—	Philadelphia and Reading Anthracite broken.	10.7	11.9	0.49 p. c.	428	10.60
	Ditto.	57.9 to 1	—	Ditto.	10.3	11.9	—	321	11.33
	Ditto.	" "	—	Geo.'s Creek Cumberland.	8.3	12.1	0.43 p. c.	397	11.99
37	Ditto.	" "	—	4 parts Geo.'s Creek Cumberland, 6 parts Anthracite Screenings.	8.7	12.2	0.49 p. c.	367	11.30
	Hor. Ret. Tub.	39.2 to 1	—	Crude Petroleum.	—	0	—	429	—
38	Hor. Ret. Tub. arranged for superheating.	40.3 to 1	33.6 to 1	Bituminous Ohio Lump.	7.6	10.9	—	501	9.35
39	Hor. Ret. Tub.	22.3 to 1	—	Anthracite Wilkesbarre broken.	9.5	20	30 deg.	394	9.55
	Hor. Ret. Tub.	" "	—	Bituminous Walston.	7.6	9.52	0.2 p. c.	445	9.48
40	Hor. Ret. Tub.	53.1 to 1	—	Geo.'s Creek Cumberland.	7.5	13.6	—	413	12.47

Summary of Tests.

Number of boiler.	Kind of boiler.	Ratio of water-heating surface to grate surface.	Ratio of steam-heating surface to grate surface.	Kind of coal	Percentage of ash.	Coal per hour per square foot of grate.	Percentage of moisture or degrees of superheating.	Temperature of escaping gases.	Water per pound of combustible from and at 212 degrees.
						Lbs.		Deg. Fahr.	Lbs.
41	Hor. Ret. Tub. double-deck.	61 to 1	3 to 1	Bituminous, Cumberland.	7	7.96	0.5 p. c.	322	11.81
42	Hor. Ret. Tub. double-deck.	65 to 1	3.3 to 1	Anthracite, Lackawanna, broken.	12.3	14.25	—	392	11.11
	Ditto.	" "	" "	Bituminous, Geo.'s Creek Cumberland.	6.7	10.97	0.5 p. c.	389	12.42
43	Hor. Ret. Tub. double-deck.	67.6 to 1	3.9 to 1	1 part Nova Scotia Culm, 3 parts Pea and Dust.	14.7	11.5	1.2 p. c.	339	10.07
	Ditto.	" "	" "	Delaware and Lackawanna Broken Anthracite.	15.5	9.4	—	335	10.39
	Ditto.	" "	" "	Geo.'s Creek Cumberland Bituminous.	6.7	8.2	—	348	10.99
44	Ditto.	62.1 to 1	4 to 1	Nova Scotia Culm. Eureka, Cumberland, Bituminous.	10.4	11.2	—	348	9.19
45	Hor. Ret. Tub. double-deck.	60.9 to 1	2 to 1	Delaware and Lackawanna Anthracite, broken.	9.7	9.3	—	375	12.03
46	Hor. Ret. Tub. double-deck, with flue heater.	58 to 1	4 to 1	Cumberland.	8.1	14.5	0.3 p. c.	373	11.00
					7.6	12.2	—	254	—

Summary of Tests.

Number of boiler.	Kind of boiler.	Ratio of water-heating surface to grate surface.	Ratio of steam-heating surface to grate surface.	Kind of coal.	Percentage of ash.	Coal per hour per square foot of grate.	Percentage of moisture or degrees of superheating.	Temperature of escaping gases.	Water per pound of combustible and at 212 degrees.
						Lbs.		Deg. Fahr.	Lbs.
47	Ditto, without flue heater.	58 to 1	4 to 1	Cumberland.	7	12.5	—	342	11.78
	Plain Cylinder.	7.5 to 1	—	Anthracite. Chestnut No. 2.	14	7.4	—	650	8.44
	" "	" "	—	Ditto.	14.7	6.1	—	567	9.22
48	Plain Cylinder.	10.9 to 1	—	Coke.	7.7	12.8	—	Above 600	6.87
	" "	" "	—	Anthracite Pea.	13.2	9.7	—	" "	7.97
49	Plain Cylinder.	10.9 to 1	—	Bituminous Cumberland.	10	7.5	—	Melts zinc	8.74
	" "	" "	—	Ditto.	10.3	7.5	—	—	8.59
	" "	" "	—	Anthracite Chestnut No. 2.	19.3	10.6	—	—	7.03
50	Galloway.	25.9 to 1	—	Geo.'s Creek Cumberland.	6.0	17.3	7.0 p. c.	533	10.00
	"	" "	—	Ditto.	7.0	21.3	0.5 p. c.	575	11.06
51	Vertical Tubular, (rolling pin).	20.6 to 1	9.3 to 1	Delaware and Lackawanna, Anthracite, broken.	8.5	17.1	—	600	8.18
	Ditto.	" "	" "	Ditto.	8.6	11.1	90 deg.	480	9.56
	Ditto.	" "	" "	Ditto.	11.7	11.0	14 deg.	434	10.07
52	Vertical Tubular with fire box.	33.5 to 1*	—	Anthracite, Lehigh egg.	8.9	15.0	—	596	10.22
	Ditto.	" "	—	Ditto.	9.4	9.1	—	446	10.51

* Total.

Summary of Tests.

Number of boiler.	Kind of boiler.	Ratio of water-heating surface to grate surface.	Ratio of steam-heating surface to grate surface.	Kind of coal.	Percentage of ash.	Coal per hour per square foot of grate.	Percentage of moisture or degrees of superheating.	Temperature of escaping gases.	Water per pound of combustible from and at 212 degrees.
						Lbs.		Deg. Fahn.	Lbs.
53	Vertical Tubular.	32.3 to 1	15.6 to 1	Anthracite Schuylkill, Broken.		15.4	42 deg.	478	10.13
	" "	" "	" "	Ditto.	11.6	24.1	59 deg.	573	9.38
	" "	" "	" "	Anthracite Schuylkill Screenings.	13.6	24.7	51 deg.	505	8.61
54	Vertical Tubular, (rolling pin).	21.1 to 1	10.7 to 1	2 parts Anthracite Screenings, 1 part Cumberland.	13.4	10.7	93 deg.	443	9.06
	Ditto.	" "	" "	Ditto.	18.8	11.6	93 deg.	434	9.27
	Ditto.	" "	" "	Geo.s Creek Cumberland.	16.7	7.3	73 deg.	413	10.27
	Ditto.	" "	" "	Anthracite Lehigh, broken.	8.2				
55	Vertical Tubular.	20.9 to 1	9.2 to 1	Ditto.	15.1	9.8	89 deg.	449	9.54
	" "	" "	" "	Ditto.	10.3	14.3	80 deg.	545	7.63
	" "	" "	" "	Ditto.	13.2	12.7	73 deg.	500	8.43
	" "	" "	" "	3 parts Screenings, 1 part Cumberland.	13.2	9.3	71 deg.	417	8.96
56	Vertical Tubular, (nest).	15.5 to 1	11 to 1	Anthracite Chestnut No. 2.	16.5	9.2	70 deg.	462	8.85
57	Vertical Tubular, (rolling pin).	19.9 to 1	11.3 to 1	Anthracite Lackawanna, egg.	12.9	7.5	31 deg.	468	8.91
	Ditto.	" "	" "	Ditto.	17	12	65 deg.	532	9.17
58	Vertical Tubular.	35.1 to 1	4 to 1	Clearfield bitumin's.	17.1	14.2	89 deg.	—	8.87
					9.3	10.3	—	423	12.29

Summary of Tests.

Number of boiler.	Kind of boiler.	Ratio of water-heating surface to grate surface.	Ratio of steam-heating surface to grate surface.	Kind of coal.	Percentage of ash.	Coal per hour per square foot of grate.	Percentage of moisture or degrees of superheating.	Temperature of escaping gases.	Water per pound of combustible and at 212 degrees.
						Lbs.		Deg. Fahr.	Lbs.
59	Vertical Tubular, (rolling-pin) with flue heater.	20.9 to 1	10.3 to 1	Geo.'s Creek Cumberland.	8	7.6	50 deg.	365	—
	Ditto. without flue heater.	" "	" "	Ditto.	8.4	9.9	52 deg.	645	8.99
60	Vert. Tub. (firebox).	44.5 to 1	15.7 to 1	Geo.'s Creek Cumberland.	7.7	13.1	18 deg.	427	12.29
61	Cast Iron Sectional.	19.7 to 1	—	Anthracite broken.	9.1	8.9	25 deg.	575	9.79
	" "	13.6 to 1	—	" "	10	8.9	153 deg.	540	9.78
62	Cast iron Sec. with flue heater.	21.2 to 1*	—	Anthracite Chestnut No. 2.	15.5	9.2	—	299	—
	Ditto without flue heater.	" "	—	Ditto.	15.2	10	—	434	9.26
63	Cast Iron Sectional.	24.2 to 1	8 to 1	Bitumin's Cambria.	12.4	9.5	29 deg.	462	9.61
64	Water Tube.	37.3 to 1	—	Anthracite Lehigh, Chestnut.	14	9.3	—	337	10.61
65	Water Tube.	40 to 1	—	Anthracite Shamokin, pea.	17.4	12.2	0.6 p. c.	353	11.44
66	Water Tube.	40.3 to 1	—	Anthracite Lehigh, broken.	9.2	17.8	—	540	9.68
67	Water Tube.	36.5 to 1	—	Anthracite Lehigh Chestnut, No. 2.	14.7	8.2	—	360	10.00
68	Water Tube with flue heater.	62.5 to 1	—	Geo.'s Creek Cumberland.	7.5	15.2	1.3 p. c.	279	—

* Below shields.

Summary of Tests.

Number of boiler.	Kind of boiler.	Ratio of water-heating surface to grate surface.	Ratio of steam-heating surface to grate surface.	Kind of coal.	Percentage of ash.	Coal per hour per square foot of grate.	Percentage of moisture or degrees of superheating.	Temperature of escaping gases.	Water per pound of combustible from and at 212 degrees.
						Lbs.		Deg. Fahr.	Lbs.
68	Ditto without flue heater.	62.5 to 1	—	Geo.'s Creek Cumberland.	7.7	16.8	—	452	10.79
	Ditto without flue heater.	" "	—	3 parts Powelton bituminous, 1 part Pea and Dust.	9.0	16.7	—	402	13.01
69	Water Tube.	31.4 to 1	4.2 to 1	Anthracite Lackawanna Chestnut No. 2.	16.4	10.9	—	428	10.36
70	Water Tube.	45.5 to 1	—	Bituminous Cambria.	10.5	16.9	0.42 p. c.	471	10.93
71	Water Tube.	48.4 to 1	—	Geo.'s Creek Cumberland.	6.4	16.3	0.4 p. c.	523	10.98

APPENDIX.

APPENDIX.

A COAL CALORIMETER.

EXPERIMENTS ON THE HEATING POWER OF VARIOUS COALS.

These experiments were made to determine the heating power of several different kinds of coal, used on evaporative tests of various boilers.

They have an important scientific application to the work of making boiler tests, and an even more important commercial application to the interests of those who use coal in large quantities. An intending purchaser may, by employing the instrument under notice, determine beforehand the exact value of the different coals offered by dealers, and may at any subsequent time determine whether he has received coal of the quality contracted for.

The apparatus used for the experiments consists of a calorimeter, in which a small quantity of the coal is burned, and the heat derived therefrom is imparted to the water which the instrument contains. The combustion is effected in an atmosphere of oxygen beneath the surface of the water, and the products of combustion are made to give up their heat to the water by mingling with the liquid itself in the act of rising to the surface and escaping. The coal used is that obtained by carefully sampling the coal employed on the evaporative test, and a representative portion of the sample is selected and finely pulverized. One gramme in weight of the pulverized coal is measured out and placed in a small platinum crucible. This is set in an inverted chamber, having a perforated base, which is immersed in the water of the calorimeter. The chamber is supplied with oxygen gas through a pipe leading from a tank containing oxygen, placed close by, and the coal is ignited. The products of combustion escape downward through the perforated bottom of the chamber, and bubble up through the water, finally being discharged at the surface of the liquid into the atmosphere. The quantity of water used is 2 kilograms, or 2000 times the weight of coal. The temperature of the water is observed before the coal is lighted, and again at the end of the process of combustion, and the rise of temperature furnishes the principal element in the required data. The material of which the instrument is composed possesses a heat capacity equivalent to that of an additional weight of 114 grammes of water. The coefficient of the apparatus for one degree rise of temperature is thus 2114. In other words, the heating power or total heat of combustion of a unit of weight of the coal is 2114 units of heat for each degree rise of temperature. If, for example, the water is heated 6 degrees Fahrenheit, the total heat of combustion of 1 pound of coal is $2114 \times 6 = 12{,}684$ B. T. U.

The results of the calorimeter tests are summarized in the following Table, and these are given in connection with the principal results of the evaporative tests of the boilers.

Test A, which gave a total heat of 13,529 thermal units, was made on a Cumberland bituminous coal, which was used in a cast-iron sectional boiler. The heating surface in the boiler amounted to 2815 square feet and the ratio of heating surface to grate surface was 53 to 1. The firing was done in a very careful manner by a special fireman. Although the temperature of the escaping gases was 460 degrees, which is excessive, the evaporative result was 10.8 pounds, uncorrected for moisture which was present in the steam to the extent of 0.5 per cent. This performance is not so high as that obtained on some of the boilers given in Part II., with the same kind of coal. The evaporation taken to represent a fair performance under favorable conditions, given on page 45 of Part I., is 11.04 pounds of water from and at 212 degrees per pound of coal.

Tests with Coal Calorimeter.

Letter designating test.	Kind of coal.	Calorimeter Tests.		Evaporative Tests.				
		Percent-age of ash.	Total heat of combustion.	Kind of boiler.	Percent-age of ash.	Coal per hour per square foot of grate.	Temperature of escaping gases.	Water per pound of coal from and at 212 degrees.
			B. T. U.			Lbs.	Deg.	Lbs.
A	Cumberland Bitum.	5.0	13,529	Sectional.	7.5	11.2	460	10.8
B	Pocahontas	4.0	14,375	Vertical Tubular.	10.0	13.6	425	9.60
C	Cumberland "	6.1	13,213	Same boiler.	8.1	14.6	400	8.93
D	Cumberland "	6.6	12,747	Locomotive.	—	—	—	—
E	Cumberland "	6.3	13,825	Locomotive.	—	—	—	—
F	Cumberland "	6.6	13,614	Locomotive.	—	—	—	—
G	Cumberland "	6.2	13,973	Water Tube.	6.1	24.9	575	10.06
H	Clearfield "	7.6	13,043	Horizontal Tubular.	10.1	14.3	—	9.77
I	Frontenac "	17.7	10,506	Horizontal Tubular.	17.3	14.7	—	6.00
J	Cape Breton "	7.8	12,515	Horizontal Tubular.	9.4	12.4	459	8.22
K	Chestnut No. 2 Anthr.	12	11,733	Horizontal Tubular.	16.0	10.5	345	9.18
L	Chestnut No. 2 "	11.8	11,373	Plain cylinder.	19.3	10.6	—	5.75
M	Cumber and Bitum.	7.6	13,867	Galloway.	7	21.3	575	10.28

NOTE.—A sample of another Pocahontas coal yielded 13,487. B. T. U. Samples of various other Cumberland coals gave, respectively, 12,874; 13,530; 12,895; 13,107; 13,318; 13,487; 13,424; 13,745.

Tests B and C were made on two coals used on evaporative tests of a plant of five vertical tubular fire box boilers encased in brick work. One of the coals was a Pocahontas bituminous, and the other a George's Creek Cumberland bituminous. The collective total heating surface amounted to 10,940 square feet. The ratio of this total surface to grate surface was 54.5 to 1, and about 70 per cent. of this was water heating surface, and 30 per cent. steam heating surface. The Pocahontas coal gave a total heat of 14,375 thermal units, and the Cumberland coal, 13,213, while the two results of the evaporative tests, including allowance for the added heat due to superheating, which was 31 degrees in the first test, and 47 degrees in the second, were respectively 9.60 pounds and 8.93 pounds. The heat given by the Cumberland coal is 8 per cent. less than that given by the Pocahontas coal, and the difference between the two evaporative results is 7 per cent. The calorimeter in these cases shows that the difference in the results of the tests is well accounted for by the difference in the quality of the fuel.

Tests D, E and F were made with three samples of George's Creek Cumberland bituminous coal, used on a series of locomotive trials. These coals were all obtained from the same dealer, and they were said to be of the same kind, but they were delivered in three different cities, and it may be presumed that they were taken from three different lots. Between the highest and lowest results of these tests there is a difference of 1078 thermal units, or about 8 per cent. of the lowest quantity, and this range indicates the variation in quality which may occur with different samples of merchantable coal, supposed to be of the same kind.

Test G was made with a George's Creek Cumberland coal, used on an evaporative test of a water tube boiler. The heating surface in the boiler amounted to 2765 square feet, and the ratio of heating surface to grate surface was 46.5 to 1. The heating power of the coal shown by the calorimeter, which was 13,973 thermal units, is higher than that given by any other Cumberland coal on the list; but the evaporative result, which was 10.06 pounds, is below that given by some of the other boilers using the same kind of coal. It is evident that the low result cannot be attributed to an inferior quality of fuel.

Test H was made on a Clearfield bituminous coal used on a test of a plant of two double-deck horizontal return tubular boilers. The area of heating surface amounted to 3860 square feet, the ratio of which to the grate surface was 67 to 1. The heating power of the coal is here 13,043 thermal units. This is 930 thermal units, or 6.6 per cent. below the result of test G made on Cumberland coal. This evidence of inferior quality furnishes a partial, if not a full, explanation of the comparatively low degree of economy shown by the evaporative test.

Test I was made on a Frontenac slack coal, mined in Southern Kansas, which was used on a test of two 72 inch horizontal return tubular boilers. The area of the heating surface was 2684 square feet, and the ratio of heating surface to grate surface 44.4 to 1. No evaporative test was made in this case, but from the results of a test of an engine, to which the boilers were furnishing steam, the evaporation appeared to be 6 pounds of water from and at 212 degrees per pound of coal. The boilers were poorly fired, and this accounts in a measure for the low economy indicated; but it is evident that this result is chiefly due to the inferior quality of the fuel. The calorimeter test gave 10,506 thermal units, which is only 75 per cent. of the result given for the best Cumberland coal in the Table.

Test J was made on a Cape Breton bituminous coal, from the Caledonia mine, which was used on a test of two 66 inch horizontal return tubular boilers. The heating surface was 1931 square feet, the ratio of which to grate surface was 29.5 to 1. A comparatively low evaporative result was obtained, being 8.22 pounds of water, and this was caused in some degree by poor firing. The low heating power shown by the calorimeter appears however to be the main cause of the inferior economy. The heat of combustion was 12,515 thermal units, or 90 per cent. of the heat shown with the best Cumberland coal on the list.

Test K was made with a Honeybrook anthracite coal of the Chestnut No. 2 size, which was used on a test of a plant of five horizontal return tubular boilers. The collective heating surface amounted to 6772 square feet, and the proportion of heating surface to grate surface was 49.3 to 1. The amount of heat shown by the calorimeter is 11,733 thermal units, and this is much below the result obtained from the Cumberland bituminous coals. The evaporative result, which was 9.18 pounds, is likewise comparatively low. Comparing these figures with those obtained on test A, the total heat of combustion of the anthracite coal is 13 per cent. less than that with the Cumberland coal, and the evaporative result is 15 per cent. less.

Test L was also made with an anthracite Chestnut No. 2 coal. In this case the boiler was of the plain cylinder type, having a heating surface of 394 square feet, and a ratio of heating surface to grate surface of 10.9 to 1. The temperature of the escaping gases was above the melting point of zinc. This boiler is the one designated as No. 49 in Part II., and the test referred to is the one numbered 99. The heating power of the coal on this test is slightly below that of test K, but this does not account in any degree for the exceedingly low evaporative result obtained, which was 5.75 pounds of water. The low result is almost wholly due to the deficiency of heating surface.

Test M was made with a George's Creek Cumberland coal, used on a test of a Galloway boiler, in which the heating surface amounted to 938 square feet, and the ratio of heating surface to grate surface was 25.9 to 1. This is the boiler which is designated in Part II., as No. 50, and the evaporative test given is the one referred to in the comments on the results of the tests in connection with that boiler. The evidence of the calorimeter test, which gave a total heat of 13,867 thermal units, is that the fuel was of superior quality to that, for example, of test A. But this did not lead to a superior evaporative result, there being on the contrary, a loss referred to that test of nearly 5 per cent. The difference in the evaporative results must be explained by the difference in the efficiency of the two types of boilers, and the data given in connection with the boiler tests show wherein this is brought about.

A UNIVERSAL STEAM CALORIMETER.

This instrument was devised by the author in 1889, and it has since that time been used where formerly the Superheating Calorimeter, referred to in Part I., was employed. It is fully described in the Transactions of the American Society of Mechanical Engineers, Volume 11, and the following account is taken from that publication. It is of simpler form than the Superheating Calorimeter, and it has a wider application, but it is no less accurate. The current of steam to be tested is first passed through a chamber in which the free moisture is deposited and measured, and subsequently it is carried through an orifice and discharged to the atmosphere, by means of which the partially dried steam is wiredrawn and superheated, and its exact final condition determined. The apparatus is shown in the following cut.

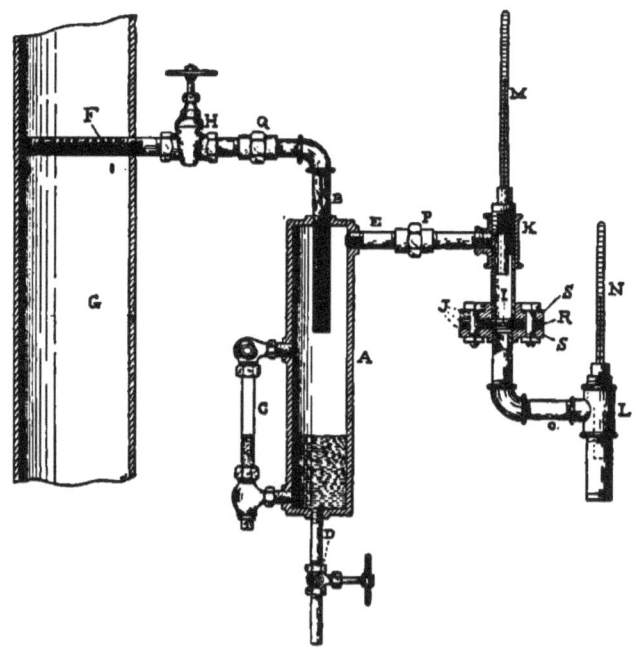

The principal parts consist of the chamber A, or "drip-box," and the wiredrawing apparatus or "heat-gauge," consisting of the orifice I, and the two thermometers M and N. The instrument is connected to the main steam-pipe G, which carries the steam to be tested, by means of the perforated pipe F, and this pipe extends across the full diameter, in order to obtain a sample of the steam tested. The orifice I opens into a pipe which is in free communication with the atmosphere. By the use of the orifice a continuous current of steam is made to pass through the whole apparatus, and the current has a constant rate so long as the pressure is constant. In the form thus far made, the supply pipe F, and the fittings, up to the drip-box are the ordinary size of one-half inch steam pipe. The drip-box is 1½ in. inside diameter and 10 inches long, and the drain-pipe D is one fourth inch pipe. The pipe leaving the drip-box is also one-half inch, and the remaining pipes and fittings are of the three-fourths inch size. The parts marked S, which enclose a plate in which the orifice is placed, are a pair of union flanges, and inserted between these flanges and under the bolt-heads are pieces of non-conducting material which prevent the direct transfer of heat through the metal walls of the high-pressure pipe above the orifice to the walls of the low-pressure pipe below it. The thermometer M rests in an oil-cup, K, as shown, and the thermometer N is arranged in a like manner.

The use of the non-conducting material shown at the points J, might seem to some unnecessary, and it may be explained that in some early experiments with the superheating calorimeter it was found absolutely necessary to cut off all solid metallic connection between the jacket and the interior heating pipe, else there would be a transfer of heat from one to the other, which would make the indications of the thermometers erroneous. The same principle applies here, and justifies the arrangement which has been adopted.

The orifice I is made about one-eight of an inch in diameter for pressures in the neighborhood of 80 pounds, and at 80 pounds pressure it discharges about 60 pounds weight of steam per hour.

The amount of moisture which the heat gauge alone will measure varies somewhat according to the pressure. If the pressure is eighty pounds, it will measure between 3 per cent. and 4 per cent. It is unnecessary to use the drip-box unless the quantity of moisture is in excess of, say, 3 per cent. The unions P and Q are therefore made interchangeable. When a test is to be made, the heat-guage is first applied directly to the union Q and a preliminary trial made, to see what the general condition of the steam is. Whenever the moisture exceeds 3 per cent., or the limiting quantity at the existing pressure, the thermometer N shows a temperature of about 213 degrees, and drops of water will generally be seen escaping from the open discharge-pipe. If the quantity of moisture is not beyond the range of the wire-drawing instrument, the temperature shown by thermometer N will be in excess of 213 degrees.

It is generally advisable to bring the complete apparatus into use, if the thermometer N shows less than 220 degrees. It is usually found, when the steam is wet enough to show this latter temperature, that the quantity of moisture varies, and thermometer N fluctuates over a considerable range, and at times it will drop to its limit of 213 degrees. Whenever, therefore, thermometer N shows a temperature of 213 degrees, either continuously or periodically, it is necessary to bring the drip-box into use.

In using the complete apparatus, the condensed water from the drip-box is drawn off, by means of the valve D, into a bucket resting on scales, and the quantity drawn off is regulated so as to keep the water level, as shown in the glass C, at a constant point. When the drip-box is used in this way, the author has found that almost the whole quantity of moisture in a sample will be deposited here, and very little moisture will be left to pass over into the heat-gauge. Indeed, the experiments show that the drip-box alone, with a suitable orifice or valve provided at the top, so as to obtain a proper circulation through it, would form a very satisfactory instrument for determining the quantity of moisture, in any case where the steam contained much of it.

When the quantity of moisture drawn off from the drain-valve D has been determined for a given time, the percentage of moisture which this represents must be found by comparing it with the total amount of steam passing through the apparatus. The total may be determined either by computation or by trial. The computation may be made by finding the exact area of the orifice, and computing the quantity which passes through by means of the formula,

$$Q = \frac{Pressure\ above\ zero \times area}{70},$$

which gives the number of pounds discharged through the orifice per second. The pressure to be used is that corresponding to the temperature shown by thermometer M. The quantity, as thus found, is accurate enough for rough comparisons. The exact quantity can be determined by conducting the steam discharged from the open end of the apparatus into a tub of water placed on scales — or, what is a better way, into a coil of lead-pipe, or iron-pipe, surrounded by flowing water, in the manner of a surface condenser, and weighing the condensed water drawn off in a given time.

A certain amount of moisture is produced by radiation from the apparatus itself, even though all the parts are well covered, as it is quite necessary that they should be, with hair felting. The readings of the instrument on the test must therefore be corrected for the loss thus occasioned. It has been the practice of the author to make these corrections by observing the indications when the apparatus is supplied with steam from the pipe G, at a time when the pressure is steady and the pipe contains nothing but dead steam, there being no current. This condition of things can generally be obtained in a factory at noon-time, when the engine is stopped, or at night, after the close of the day's work. It may fairly be presumed that the apparatus is then supplied with dry steam and whatever moisture collects in the drip-box A, and whatever difference is shown by thermometers M and N, is due simply to the loss of heat from radiation. When the loss from radiation has been thus obtained, the quantity representing that due to the drip-box is simply subtracted from the weight of water drawn off during the same length of time on the main test. The way in which the correction is applied to the readings of thermometers M and N is to take the reading of thermometer N on the radiation test when thermometer M indicates an average, and use this reading as a starting-point. The indication of thermometer N on the main test is then simply subtracted from this normal reading. For example, suppose the average reading of the upper thermometer (M) during the main test is 312 degrees; suppose the lower thermometer (N) indicates an average of 260 degrees on the main test, and on the radiation trial suppose it indicates 267 degrees when thermometer M shows 312 degrees; the process would be simply to subtract 260 from 267, and this would give seven degrees as the cooling effect produced by the moist steam discharged on the main test.

In order to compute the amount of moisture from the loss of temperature shown by the heat-gauge, the number of degrees of cooling of the lower thermometer (N) is divided by a certain coefficient, representing the number of degrees of cooling due to 1 per cent. of moisture. This coefficient depends upon the specific heat of superheated steam, which, according to Regnault's experiments, is 0.48. In other words, the heat represented by 1 degree of superheating is 0.48 of a thermal unit. The author's experiments show that this quantity cannot be applied exactly to the form of instrument under consideration. The quantity to be used varies somewhat according to the degree of moisture. For an instrument working under a temperature of 314 degrees, by the upper thermometer, and with a cooling by the lower thermometer from 268 degrees to 241 degrees, the quantity was found to be about 0.42. When the cooling, however, was from 266 degrees to 225 degrees, the quantity to be used was found to be about 0.51. The experiments have not as yet covered a sufficient range to determine the exact law which can be applied to every case, but it seems probable that the specific heat is more or less constant until the temperature by the lower thermometer approaches the point of saturation for the low-pressure steam, while beyond this point the specific heat rapidly increases. For the present, it is assumed that the quantity 0.42 is the proper one to apply whenever the temperature by the lower thermometer is above 235 degrees, and that in cases where the temperature is below 235 degrees, the quantity to be used is an increasing one, reaching perhaps to 0.55 when the temperature drops to 220 degrees.

One per cent. of moisture, now, represents the quantity of heat determined by multiplying the latent heat of one pound of steam, having a pressure corresponding to the indication of thermometer M, by 0.01, and this product is to be divided by 0.42 (provided the lower temperature is not below 235 degrees), in order to express it in terms of degrees of superheat. For example: When thermometer M shows 312 degrees, the latent heat is 894 thermal units, and 1 per cent. of this is 8.94; dividing by 0.42, the number of degrees of superheat corresponding to 1 per cent. of moisture is found to be 21.3. For several other temperatures, which cover the ordinary range that would commonly be used, the necessary coefficient is given in the following table:

Temperature by Thermometer M.	Coefficient.	Temperature by Thermometer M.	Coefficient.
270	22.0	320	21.1
280	21.8	330	21.0
290	21.7	340	20.8
300	21.5	350	20.6
310	21.3	360	20.5

The utility of the instrument, and the general manner in which it operates in practice, may best be shown by reference to the various tables which are appended, giving the results of a number of tests on different boilers, made by the author.

TEST N.

The first test was made when using simply the wire-drawing part of the instrument, or "heat-gauge." Only one thermometer was applied — that is, the lower thermometer — and the temperature of the steam above the orifice is only to be learned by the indications of the pressure-gauge attached to the boiler. The boiler was one which gave steam of varying degrees of dryness, and it was admirably adapted for a test of this kind. It consisted of two shells, the lower one of which was nearly filled with tubes, and the upper one served the purpose of a drum. The connection between the two shells was by a single neck at the front end. Two steam-pipes carried away the supply of steam, one being attached at the front end of the drum, directly over the connecting neck, and the other being attached to the rear end. The smoke and products of combustion, on leaving the tubes, passed over the exterior surface of the drum on their way to the chimney, and the drum thus furnished a certain amount of steam-heating surface. The calorimeter was attached first to one of these pipes and then to the other. The boilers at this place were two in number, and the calorimeter was attached to only one. When the instrument was applied to the front pipe, the quality of the steam indicated was quite variable, but when it was attached to the rear pipe, the quality became nearly constant, and the indications pointed to a small amount of superheating. The indications of the instrument on the front pipe revealed an exceedingly interesting state of affairs. The quality of the steam fluctuated periodically over a considerable range, varying from a condition of extreme wetness to a nearly dry condition. When the indications pointed to the largest amount of moisture, the thermometer showed about 213 degrees, and drops of water issued with the steam at the point of discharge. The periodical fluctuations were surprising, until a careful observation of the times when they occurred showed that the wet steam was produced according to the rate of production of steam in the boiler to which the calorimeter was attached. The firing of the two boilers was done alternately, and when the boiler to which the calorimeter was attached was fired, there was a cessation in the generation of steam in this boiler, and the other boiler in which the fire was active was drawn upon to make up the deficiency. At this time the thermometer in the calorimeter boiler always showed increasing indications. Whenever, on the contrary, the other boiler was fired, and at times when the fresh fires in the calorimeter boiler had become active again, the indications of the thermometer would rapidly fall. At such times the auxiliary boiler became for the moment inoperative, and the calorimeter boiler was called upon to furnish the greater part of the steam. The lower shell of the boiler being nearly filled with tubes, and the generating surface in this shell being very small, there was an active tendency for this shell to carry up into the drum a mixture of water and steam, and this would easily find its way into the front pipe, directly above, whenever the boiler was doing much work. The alternate heating and cooling, due to the periodical firing, produced alternately very wet and nearly dry steam in the front part of the drum, and, as a consequence, it produced a varying degree of moisture in the steam which passed through the front pipe, as noted.

Dimensions and Other Data regarding Boiler used on Test N.

1. Diameter of main shell, ... 54 in.
2. Length of main shell and length of tubes, . . . 17 ft.
3. Number of tubes 4 inches outside diameter, ... 45
4. Size of grate, ... 4x8 ft.
5. Area of grate surface, ... 32 sq. ft.
6. Area of water-heating surface, ...934 sq. ft.
7. Area of steam-heating surface, ...107 sq. ft.
8. Area of total heating surface, ... 1,041 sq. ft.
9. Collective area through tubes, ... 3.4 sq. ft.
10. Ratio of total heating surface to grate surface, . . 32.5 to 1
11. Ratio of grate surface to tube area, ... 9.3 to 1
12. Kind of coal used, ... Schuylkill pea.
13. Percentage of ashes, ... 17.8 per ct.
14. Coal burned per hour per square foot of grate, . . 11.5 lbs.
15. Water evaporated per square foot of heating surface per hour, ... 3.1 lbs.
16. Average temperature of flue gases, ...420 deg.
17. Average draught suction, ... 0.24 in.
18. Water evaporated per pound of combustible from and at 212 degrees, uncorrected for moisture, . . . 11.07 lbs.

TABLE No. 1.

Double-deck Horizontal Return Tubular Boiler, Calorimeter attached to Front Steam-pipe.

Time.	Boiler Gauge.	Temperature shown by Lower Thermometer of Calorimeter.	Condition of Steam at outlet of Calorimeter as it appeared to the eye.	Position of Damper.	TIME OF FIRING.	
					Auxiliary Boiler.	Boiler to which Calorimeter was applied.
10.00	–	236	Dry	Wide open	10.00	–
10.01	–	226	Wet*	"	–	–
10.02	–	242	Dry	"	–	10.02
10.03	–	256	"	"	–	–
10.04	–	264	"	"	–	–
10.05	–	269	"	"	–	–
10.06	–	271.5	"	"	–	–

* By the term "wet" is meant that drops of water emerged from the outlet of the calorimeter.

264 BOILER TESTS.

TABLE NO. 1n. — Continued.

Time.	Boiler Gauge.	Temperature shown by Lower Thermometer of Calorimeter.	Condition of Steam at outlet of Calorimeter as it appeared to the eye.	Position of Damper.	TIME OF FIRING.	
					Auxiliary Boiler.	Boiler to which Calorimeter was applied.
10.07	–	273.5	"	½ open.	–	–
10.08	–	272	"	"	–	–
10.09	–	265.5	"	"	–	–
10.10	–	264.5	"	"	–	–
10.11	–	230	"	"	10.11½	–
10.12	–	218	Wet	Wide open	–	–
10.13	–	222	Dry	"	–	–
10.14	–	241	"	"	–	–
10.15	–	256	"	½ open	–	10.15
10.16	–	268	"	"	–	–
10.20	88	258	"	½ open	–	–
10.21	–	246	"	"	–	–
10.22	–	226	"	"	–	–
10.23	–	220.5	Wet	½ open	10.22¾	–
10.24	–	215.5	"	Wide open	–	–
10.25	–	230	"	"	–	–
10.26	–	246.5	Dry	"	–	–
10.27	–	256.5	"	"	–	10.27
10.28	88	264	"	"	–	–
10.29	–	269	"	½ open	–	–
10.30	–	270	"	"	–	–
10.31	–	262	"	"	–	–
10.32	–	241	"	"	–	–
10.33	–	221	"	"	10.33	–
10.34	–	214.5	Wet	"	–	–
10.35	–	225	"	Wide open	–	–
10.36	–	242	Dry	"	–	–
10.38	87	261	"	"	–	–
10.39	–	257.5	"	½ open	–	–
10.40	–	255.5	"	"	–	–
10.41	–	259	"	"	–	10.41
10.42	–	265	"	"	–	–
10.43	–	269.5	"	"	–	–
10.44	–	272	"	"	–	–
10.45	–	274	"	"	–	–
10.46	–	273	"	Wide open	–	–
10.47	–	253	"	"	–	–
10.48	–	227	"	½ open	–	–
10.49	–	223.5	"	"	–	–
10.50	–	235	"	"	–	–
10.51	–	241.5	"	"	10.51	–
10.52	–	236	"	"	–	–
10.53	–	237	"	Wide open	–	–
10.54	87	245	"	"	–	–
10.55	–	251.5	"	½ open	–	10.55¾
10.56	–	258	"	"	–	–
10.57	–	263	"	Wide open	–	–
10.58	–	267.5	"	"	–	–
10.59	–	265	"	"	–	–
11.00	–	235	Wet	"	11.00	–
Normal.	85	288 (Rear pipe.)		–	–	–

In Table No. 1n, the indications of the thermometer, as also data regarding the condition of the issuing steam as it appeared to the eye, the position of the damper, and the time of firing of each boiler, are given for nearly every minute during an hour's test. The fluctuating character of the readings, as influenced by the time of firing, is clearly shown in this record. Notice, for example, the reading at 10.27, when the boiler to which the calorimeter was applied was fired with fresh coal. The reading is 256.5°, and, compared with the previous readings, the indication of the thermometer is rapidly rising. At 10.28 it reaches 264°, and at 10.30 it is 270°. At this time — three minutes after firing — it is evident that the boiler began to recover its normal rate of production, and the indications of the thermometer began to fall. At 10.33 they had dropped to 221°, and at the same time the auxiliary boiler was fired, which occurrence threw nearly the whole work of production upon the calorimeter-boiler. This was followed by the thermometer going down to 214.5°, and the issuing steam assuming a wet appearance to the eye. The thermometer then began to rise, and at 10.38, five minutes afterward, it had reached 261°. It is presumed that the fire in the auxiliary boiler had now become quite active, while that in the calorimeter-boiler was somewhat cooled.

The normal reading of the instrument was not determined when applied to the front pipe, but, taking the indication for the normal as determined for the rear pipe, which was 288°, the cooling effect due to moisture, for the best indication — viz., 274° at 10.45, is $288° - 274° = 14°$; and this, divided by the proper coefficient, viz., 21., gives for the percentage of moisture, under these circumstances, 0.66. The lowest indication, viz., 214.5° at 10.34, shows a cooling effect due to moisture of $288° - 214.5° = 73.5°$; and this, divided by the assumed coefficient for this temperature, which is 16.9, gives, for the percentage of moisture, 4.34 per cent. This percentage, however, does not show the whole of the moisture, because the range of the instrument was evidently exceeded.

The record of the test when the calorimeter was applied to

the rear pipe is given in Table No. 2n. The indications of the thermometer vary from 288° to 297.5°, with a normal reading of 288° at 85 pounds pressure. There is continual evidence here of superheating, and this might be expected from the fact of the steam-heating surface, of which the steam issuing from this end of the drum had the benefit.

TABLE No. 2n.

Double-Deck Horizontal Return Tubular Boiler (same as preceding): Calorimeter Attached to Rear Steam-Pipe.

Time.	Boiler Gauge.	Temperature shown by lower Thermometer of Calorimeter
1.00	84	291.5
1.01	82.5	290.5
1.07	72	288
1.11	73	287.5
1.19	69	289.5
1.23	71	289
1.49	82	289.5
1.54	88	292
1.58	87.5	294.5
2.02	84	295.5
2.04	82	295
2.09	81	295.5
2.12	84	296.5
2.15	86	297.5
2.17	86.5	297.5
2.23	85	296
2.27	87	297.5
Normal	85	288

TEST O.

Test O was made on a water tube boiler. On this test the heat-gauge only was in use. Appended is a table showing the general dimensions of the boiler, and the conditions under which it was operated.

Dimensions and other Data regarding Boiler referred to in Test O.

1. Number of sections. 14
2. Number of tubes, 4 in. outside diameter, in each section. . 9
3. Total number of tubes. 126
4. Diameter of drum. . . . . 36 in.
5. Size of grate. . . . . 7 x 8.5 ft.
6. Area of heating surface. . . . . . . 2,765 sq. ft.
7. Area of grate surface. . . . . 59.5 "

APPENDIX. 267

   8. Ratio of heating surface to grate surface. . . . . 46.5 to 1
   9. Kind of coal used. . . . . George's Creek, Cumberland
10. Percentage of ashes. . . . . . . 6.3 per ct.
11. Coal consumed per hour per square foot of grate. . . 17.9 lbs.
12. Water evaporated per square foot of heating surface from
      100 degs. at 70 lbs. pressure per hour. . . . . 3.38 lbs.
13. Average temperature of flue gases. . . . . . 545 degs.
14. Average draught suction. . . . . . . 0.49. in.
15. Water per pound of combustible from and at 212 degs. . 10.42 lbs.

On this test, readings of the instrument were taken at intervals of from one to five minutes during most of a ten hours' run, and the full set of observations is given in Table No. 3o. Remarks are given, in connection with many of the readings, as to the condition of the fire, height of water in the gauge-glass, and other information. It will be seen from this table that the quantity of moisture in the steam was quite variable, though never excessive. The smallest indication of the lower thermometer was that taken at 4.56 P.M., when the reading was 249°, and the highest indication was at 7.02 A.M., when the reading was 276°. The range corresponds to a little over 1 per cent. of moisture. Using the normal of 280° for a temperature of 331° by the upper thermometer, which was found at a time when the boiler was discharging little, if any, steam, the lowest reading of 249°, gives a cooling effect due to moisture of 280° — 249° = 31°, and the highest reading gives 280° — 276° = 1° for the cooling effect of moisture. The coefficient for 330° is 21° — that is, the cooling due to 1 per cent. of moisture. The two extreme percentages of moisture are, therefore, $\frac{31}{21} = 1.48$ per cent., and $\frac{1}{21} = 0.048$ per cent.

The variations in the indications of the lower thermometer were so marked, and occasionally so rapid, that an attempt was made to ascertain whether these variations could be accounted for by any differences in the condition of the fire, the height of water, or the manner of feeding the water. At one time it was thought that the lowering of the thermometer was caused by an increased activity of the fire. Notice the reading at 8.42½ A.M., which was 274°, and the next reading, which fell to 251°, 7½ minutes afterward, and between these two readings the fire was shoved back, and new coal added. The next time, however, that the fire was shoved back, which was at 9.12 A.M., there was no immediate change in the indication of the thermometer, though at 9.20 the reading had fallen to 259°, and this fall may finally have been due to the increased activity of the fire. A little further on, at 9.55, the reading was 2 77°. Shortly afterward the fire was shoved back, and immediately the temperature fell to 257°, and two minutes later to 253°. Take the reading, however, at 12.28, when the fire was treated in the same manner — there was no fall in the temperature, even after slicing the fire, and even when the water was pumped up to quite a high point.

TABLE NO. 3 o.

Water Tube Boiler.

Time.	Upper Thermometer.	LOWER THERMOMETER. Normal 279 with Upper Thermometer 330. Say 280 at 331.	Remarks as to Height of Water, State of Fire and other Observations.
6.33	329	277	
7.02	331	279	
7.35	331	271	
8.12	331	255	
8.20	331	272	
8.22½	331	276	
8.30	331	273	
8.32½	331	276	
8.40	331	274	
8.42½	331	274	Shove back and fire at 8.46.
8.50	330	251	
8.52½	330	252	
8.55	331	264	
9.00	329	275	
9.05	330	271	
9 10	330	277	
9.13	330	277	Shove back at 9.12.
9.14	331	275	Firing; water, 3 inches.
9.15	331	276	
9.20	331	259	
9.22½	331	251	Water 4½.
9.25	331	265	
9.31	331	275	Shove back and fire, 9.32–33.

TABLE NO. 3 o. (Continued.)

Time.	Upper Thermometer.	LOWER THERMOMETER. Normal 279 with Upper Thermometer 330. Say 280 at 331.	Remarks as to Height of Water, State of Fire, and other Observations.
9 34½	330	275	Water, 5½.
9.35	330	271	
9.37	331	268	
9.40	331	275	
9.45	331	276	
9.50	331	276	Front fired; water 4 inches.
9.55	331	277	Water, 3 inches.
10.01½	331	257	Shove back and fire, 9.58–10.00.
10.03½	331	258	Water, 4 inches; feeding fast.
10.09	331	264	
10.12	331	275	
10.17	330	277	
10.22	331	271	Front shoved back and fired.
10.28	330	273	Water, 5½.
10.32	331	252	" 4½.
10.35	331	271	
10.41	330	275	
10.45	330	260	Water, 6½.
10.50	331	256	Front shoved back and fired.
10.55	331	250	Water, 6; pump slow.
11.07	330	263	Feeding fast; height, 4 inches.
11.15	331	260	
11.15½	331	251	Height, 5½.
11.18	330	256	Front fired, height, 6.
11.20½	–	263	
11.22	331	258	
11.23	331	253	
11.32	331	259	
11.35	331	274	Front fired; height, 6.
11.38	330	276	
11.41	330	275	
11.43	331	256	Water, 7½.
11.45	331	261	Front shoved back.
11.47	331	267	
11.50	331	260	Height, 6½.
11.55	331	262	" 6 .
12.00	331	253	" 7.
12.06	332	277	Height, 7; damper shut.
12.10	332	259	" 7; " open.
12.15	332	260	" 6½.
12.22	332	252	" 5.
12.28	331	275	Front shoved back and firing.
12.31	330	277	Height, 5.
12 34	329	275	
12.38	328	275	Front sliced; height, 7.
12.40	327	271	
12 45	327	273	
12.51	329	275	

TABLE NO. 3 o. (Concluded.)

Time.	Upper Thermometer.	Lower Thermometer. Normal 279 with Upper Thermometer 330. Say 280 at 331.	Remarks as to Height of Water, State of Fire, and other Observations.
12.59	330	258	Height, 7.
1.13	330	275	" 7.
1.21	330	276	" 8.
1.30	330	277	" 4.
1.38	330	261	Feeding fast; height 5.
1.43	329	262	Front shoved back and fired.
1.50	330	264	Feeding fast; height 6.
1.57	330	256	Height, 5.5.
2.08	331	266	" 5.5.
2.14	331	253	" 5.2.
2.20	330	273	Shoved back, 2.21.
2.22	329	268	Fired, 2.22½.
2.23	329	261	
2.24	329	259	
2.25	329	252	
2.26½	329	252	Height, 6.5.
2.28	329	255	Slow down pump, 2.29½.
2.30	330	269	
2.35	331	276	
2.46	331	261	Height, 3.
2.52	331	263	Shove back, 2.50.
2.54	330	256	
3.04	331	266	Height, 3.
3.09	330	276	" 3.
3.20	331	255	" 2.5.
3.25	331	255	" 3.
3.32	331	273	" 4.
3.40	331	275	Shoved back and fired, 3.35; height, 6.5.
3.45	331	277	Height, 5.
3.53	331	268	Height, 3.5.
3.59	331	270	" 4.5.
4.06	331	264	" 3.5.
4.11	331	271	Shoved back, 4.10; fired, 4.11½.
4.12	331	270	Height, 3.5. (?)
4.13	331	271	
4.14	331	266	Height, 5.5.
4.15	331	262	" 6.
4.17	330	273	
4.24	331	265	Height, 4.5.
4.28	330	273	" 4.
4.39	326	275	" 4.
4.45	326	275	" 4.5.
4.53	330	258	Front shoved back and fired, height, 5.5.
4.56	330	249	Height, 6.5.
5.04	330	276	" 5.5.
5 16	331	276	
5.22	331	275	Height, 5.5.
5.28	333	276	Damper shut.
5 33	331	278	Height, 5.4.

TEST P.

Test *P* was made on a plant of two horizontal return tubular boilers, and in this case the complete form of the instrument was used. The full set of observations is given in Table No. 4*p*. Considerable interest attaches to this test on account of the priming of one of the boilers, which was brought about by design. This boiler was filled with water to nearly the top of the glass, and the bituminous coal fire was barred up, the damper opened wide, and the boiler made to do its maximum amount of work. The time when this occured was 2.05 P.M., and shortly afterward the quantity of water collecting in the drip-box began to increase, and for the seven and a half minutes between 2.15 and 2.22½, 18.9 ounces of water were withdrawn. This represents 9.28 pounds per hour, or about 18.5 per cent. of moisture, the quantity of steam used being estimated at 50 pounds per hour.

Taking the ordinary indications of the instrument during the 20 minutes' time between 12.55 and 1.15 P.M., the drip-box discharged 2.7 ounces, or 0.51 of a pound per hour. The average reading of the upper thermometer was 304.9°; and of the lower thermometer, 268.2°. Comparing these observations with the normal readings, it is seen that there was but a trifling indication of moisture.

It may be added that the priming was found to be due, when the water was carried too high, to the presence of vegetable matter in the water, and to too infrequent blowing off.

TABLE No. 4p.
Two Horizontal Return Tubular Boilers, Complete Calorimeter in Use.

Time.	Upper Thermometer.	Lower Thermometer.	Height of Water in Glass in Sixteenths of an inch.	Remarks.
12.22½	306	265	3	Practically no
12.25	304.5	267	5	steam drawn off
12.27½	302	267	7 scant	from boilers
12.30	301	267	8 sca t	between 12.25½ and
12.32½	303	267	10	12.47½
12.35	305	267	13 +	
12.37½	303	268	15 +	
12.40	304	268.5	17 scant	
12.42½	304	268.5	18 +	Draw off from
12.45	303	268	20	drip-box, at 12.46,
12 47,	302	267	2 +	2.7 ounces.

TABLE No. 4 p. (Continued.)

Time.	Upper Thermometer.	Lower Thermometer.	Height of Water in Glass in Sixteenths of an inch	Remarks.
Average normal readings.	303.6	267.3	.39 lb. per hour	
12.50	302	267	4 +	
12.52½	302	267	6 scant	
12.55	303	267	8 +	Engine started
12.57½	303	267	11	at 12.55.
1.00	303	267	14	
1.02½	305	268	16 +	
1.05	306	268	19	
1.07½	307	269	21 +	
1.10	307	269	24	
1.12½	306	270	7	Draw off from
1.15	304	269	8 +	drip-box, at 1.11,
1.17½	302	268	9 +	2.7 ounces.
1.20	301	268	10 +	
1.22½	302	267	12 +	
1.25	302	267	14 +	
1.27½	302	267	16 +	
1.30	304	267	19	
1.32½	304	268	22	
1.35	304	268	23	Draw off from
1.37½	303	268	5.5	drip-box, at
1.40	301	268	6 +	1.36, 2.7 ounces.
1.42½	300	267	8	
1.45	298	266	10	Water being
1.47½	298	266	10 +	pumped to a high
1.50	300	265	13	point.
1.52½	302	265	16	Draw off from
1.55	304	266	19	drip-box, at 2.01,
1.57½	306	267	22	2.7 ounces.
2.00	306.5	268	24	At 2.05 damper
2.02½	307	269	8	of one boiler
2.05	307	270	11	shut. Fire in
2.07½	308	270	22 +	the other boiler
2.10	306	270	29	barred up and
2.12½	308	270	12	water at a high
2.15	307	270	14	point.
2.22½	306	255	12	Draw off from
2.25	308	266	15	the drip-box, at
2.27½	310	270	18 +	2.11, 2.7 ounces.
2.30	310	270	20 +	Between 2.15 and
2.32½	310	270	21 +	2.22 draw off from drip-box 18.9 ounces.

TABLE No. 4 p. (Concluded.)

Time.	Upper Thermometer.	Lower Thermometer.	Height of Water in Glass in Sixteenths of an inch.	Remarks.
2.35	309	271	23	At 2.25 height of water in the active boiler had fallen 4 inches. Draw off from drip-box, at 2.36, 2.7 ounces. At 2.42½ dampers of both boilers open.
2.37½	310	271	7	
2.42½	301	268	8	
2.45	298	267	8 +	
2.47½	299	266	10	
2.50	301	266	12	
2.52½	305	267	17	
2.55	308	266	21	
2.57½	310	269	23	
3.00	312	270	24 +	

INDEX.

Air above fuel, Boilers which admit, 81, 89, 111, 123, 132, 157, 192, 217, 224.
Air above fuel, Prevention of smoke by admitting, 54, 62, 113, 137.
Air and steam through cast iron globes at bridge wall, Economy of admitting, 89, 179.
Air at bridge wall, Economy of admitting, 52, 111.
Air leakage through unsound brick work, 201, 215, 232.
Air over fuel, Economy of admitting, 52, 53, 54, 55, 56, 83, 91, 111, 125, 133, 195.
Air space in grates, Tests with different proportions of, 60, 95.
Air through bridge wall and through side walls of furnace, Economy of admitting, 55, 83, 125, 133, 195.
Air through bridge wall and through supplementary wall behind it, Economy of admitting, 54, 158.
Allowance for superheating, 22, 186.
Alternate firing, Boiler with two furnaces for, 38, 121.
Anthracite coal, Tests with, 67, 73, 78, 89, 92, 96, 98, 102, 104, 107, 113, 116, 119, 126, 127, 129, 131, 146, 149, 154, 162, 165, 169, 174, 177, 179, 183, 187, 190, 192, 196, 199, 201, 211, 213, 217, 219, 221, 222, 228.
Artificial draught with blower, Economy of, 61, 87.
Ash in different coals, Percentage of, 111, 164.
Automatic vs. hand regulation of draught, 60, 94.

Banking fires, Loss from, 62, 179.
Barrel Calorimeter, 17.
Bituminous coal, Tests with, 70, 78, 89, 101, 110, 121, 132, 134, 136, 139, 140, 145, 148, 150, 152, 156, 157, 159, 163, 166, 168, 171, 179, 180, 194, 203, 205, 208, 215, 224, 230, 231.
Blower draught, Boilers with, 70, 75, 86, 190.
Boiler setting, 51.
Boilers, Cast iron Sectional, 211, 213, 215.
Boilers, Direct Tubular, 86, 105, 149.
Boilers, double-deck, 159 to 171.
Boilers, Galloway, 180.
Boilers, Horizontal Return Tubular, 67 to 157.
Boilers, Plain Cylinder, 174 to 180.
Boilers, Plant of two or more, See Plant.

(274)

INDEX. 275

Boilers, Vertical Tubular, 183 to 210.
Boilers, Water Tube, 217 to 233.
Boilers with miscellaneous details of construction, See Air, Alternate firing, Blower, Detached furnace, Double passage, Fire box, Flue heater, Superheating, Water leg.
Brick arch in furnace, Boiler with, 149.
Bridge wall, Effect of shape of, upon economy, 51.
Broken coal, Tests with Anthracite, 73, 78, 101, 107, 118, 121, 131, 148, 150, 163, 166, 169, 183, 194, 197.

Calorimeter, Barrel, 17.
Calorimeter for determining heat of combustion of coal, Appendix.
Calorimeter, Superheating, 18, 19.
Calorimeter tests of the quality of steam, 62.
Calorimeter, Universal steam, Appendix.
Cast Iron Sectional Boilers, 211, 213, 215.
Cast Iron Sectional Boilers, Comparative economy of, 40.
Cast Iron Sectional Boilers with flue heater, 213.
Chestnut coal, Tests with, 67, 102, 104, 107, 217.
Chestnut No. 2 coal, Economy of, 49.
Chestnut No. 2 coal, 115, 127, 174, 179, 199, 213.
Coal, Cambria bituminous, 215, 230.
Coal, Clearfield bituminous, 203.
Coal, Cumberland, 70, 79, 91, 101, 110, 121, 122, 136, 139, 148, 150, 157, 159, 163, 166, 168, 171, 179, 180, 194, 205, 208, 224.
Coal, Delaware and Lackawanna Broken, 169.
Coal, Franklin, 81.
Coal, Hazelton Chestnut, 67.
Coal, Honeybrook Lehigh Chestnut, pea and broken, 107.
Coal, Kalmia, 81.
Coal, Lackawanna Broken, 73, 79, 101, 163, 166, 183,
Coal, Lackawanna Chestnut No 2., 228.
Coal, Lehigh Broken, 101, 121, 194, 197, 221.
Coal, Lehigh Chestnut, 102, 104, 107, 217.
Coal, Lehigh Chestnut No. 2, 222.
Coal, Lehigh Egg, 91, 126, 187.
Coal, Mixture of Buckwheat and Clearfield, 142.
Coal, Mixture of Pea and Dust and Clearfield, 115.
Coal, Mixture of Pea and Dust and Cumberland, 79, 86, 110, 123.
Coal, Mixture of Pea and Dust and Nova Scotia Culm, 79, 118, 166.
Coal, Mixture of Pea and Dust and Powelton, 227.
Coal, Mixture of Screenings and Cumberland, 91, 101, 150, 194, 197.
Coal, Mixture of Screenings and Nova Scotia Culm, 75, 81.
Coal, Nova Scotia Culm, 132, 134, 166.
Coal, Ohio Lump, 50, 152.
Coal, Philadelphia and Reading Broken, 148, 150.
Coal, Schuylkill Broken, 191.
Coal screenings, 191.

Coal, Shamokin Pea, 219.
Coal, Walston bituminous, 49, 145, 156.
Coal, White Ash Broken, 118, 131.
Coal, Wilkesbarre, 154.
Coke, Boilers using, and economy of same, 50, 139, 178.
Combustion chamber or space behind bridge wall, Arrangement of, 51.
Commercial tests, Engineering tests and, 11, 12.
Comparison between boiler with air admission, and boiler without air admission, above fuel, 52, 123, 132.
Comparison between boiler with three inch tubes, and boiler with three and a half inch tubes, 35, 134.
Comparison between different kinds of boilers, 36.
Comparison between double-deck boilers and ordinary horizontal tubular boilers, 39, 162.
Comparison between economy of different kinds of fuel, 43, 109, 113, 116, 149, 162, 196.
Comparison between saturated steam boilers and superheated steam boilers, 22.
Comparison between vertical boilers and horizontal tubular boilers, 23, 186.
Comparative draught required for different fuels, 48, 49, 119.
Comparative labor of firing different fuels, 80, 119.
Computing results, Method of, 19, 20, 21.
Conducting ideal tests, Method of, 12.
Conducting tests given in Part II, Method of, 13, 14, 15.

Detached furnace, Boilers with, 36, 86, 149.
Direct tubular boiler with common furnace, 104.
Direct tubular boilers with detached furnace, 86, 149.
Direct tubular boilers with detached furnace, Comparative economy of, 37.
Double-deck boilers, 159 to 171.
Double-deck boilers, Comparative economy of, 39 163.
Double furnace for alternate firing, Hor. Ret. Tub. boiler with, 38, 121.
Double passage for products of combustion, Boilers with, 75. 98, 154.
Draught with blower, Economy of artificial, 61, 88.
Draught, Automatic vs. hand regulation of, 60, 94.
Draught, Effect of flue heater on, 58.
Draught for vertical boilers, 195, 202.
Draught gauge, 16.
Draught, Method of taking, 16.

Economical temperature of escaping gases with anthracite and bituminous coal, 28, 31.
Economy of carrying high water in vertical boiler, 187.
Economy of flue heaters, 56, 145, 174, 207, 226.
Economy of wetting bituminous coal, 61.
Effect of admitting air above fuel in preventing smoke, 53, 62, 113, 137, 159.
Effect of admitting air above fuel upon flue temperature, 113.
Effect of admission of air through unsound brick work upon draught and economy, 26, 201, 215, 232.

INDEX. 277

Effect of flue heater on draught, 58.
Effect of flue temperature on economy, 26, 28.
Effect of proportion of tube area to grate surface upon economy, 34.
Effect of size of shell upon economy, 34.
Effect of size of tubes upon economy, 35.
Engineering tests and commercial tests, 11, 12.
Escaping gases, Economical temperature of, with Anthracite and bituminous coal, 28.

Feed water heated by coil of pipe in flue, 76, 221.
Feed water, Method of weighing, 15.
Fire box vertical boilers, 27, 187, 203, 208.
Flue heater, Boilers with, 142, 171, 205, 213, 224.
Flue heaters, Economy of, 56, 145, 174, 207, 226.
Flue heater with cast-iron sectional boilers, 213.
Flue heater with horizontal tubular boilers, 142, 171.
Flue heater with vertical tubular boilers, 205.
Flue heater with water tube boilers, 224.
Flue temperature, Effect of, upon economy, 28.
Forced draught, Boilers provided with, 70, 75, 86, 190.
Free burning coal, 145, 152.
Furnace for burning bituminous coal without smoke, 136.

Galloway boiler, 180.
Galloway boiler, Comparative economy of, 42.
Gases, Economical temperature of, with Anthracite and bituminous coal, 28.
General conditions under which tests were made, 12, 13.
General conditions which secure economy, 28.
Grates with 50 per cent. air space, vs. grates with 60 per cent. air space, Economy of, 95.

Hand regulation vs. automatic regulation of draught, 60, 94.
Heated air above fuel, Boiler supplied with, 84.
Heating feed water by coil of pipe in flue, 76, 221.
High flue temperature in cast-iron sectional boilers, plain cylinder boilers, and Galloway boiler, 31.
High flue temperature in vertical boilers, 26.
High water in vertical boiler, Economy of, 187.
Horizontal direct tubular boiler with common furnace, 105.
Horizontal direct tubular boiler with detached furnace, 86, 149.
Horizontal return tubular boiler, 42 inch, 154, 156.
Horizontal return tubular boiler, 48 inch, 67, 78, 92, 96, 102, 104, 129, 136.
Horizontal return tubular boiler, 54 inch, 81, 145.
Horizontal return tubular boiler, 60 inch, 113, 123, 131, 157.
Horizontal return tubular boiler, 66 inch, 138.
Horizontal return tubular boiler, 72 inch, 127, 134, 152.
Horizontal return tubular boiler with additional pipe surface in furnace, 67, 142.

Horizontal return tubular boiler with double passage for products or combustion, 38, 75, 98, 134.
Horizontal return tubular boiler with superheater, 67, 154.
Horizontal return tubular boiler with water leg front, 89, 107.

Influence of flue heater on the draught, 59.
Interests which led to tests, 9.

Labor of firing with different fuels, Comparative, 80, 119, 196.
Leakage of air through unsound brick work, 26, 201, 215, 232.
Location of boilers tested, 10.
Loss by air leakage through unsound brick work, 26, 201, 215, 232.
Loss by banking fire, 62, 179.
Loss of heat in vertical boilers at chimney, 26.

Meaning of lines given in tables, 19.
Method of computing results, 20.
Method of conducting ideal tests, 12.
Method of conducting tests given in Part II, 13.
Method of determining quality of steam, 17.
Method of taking temperature and draught suction, 16.
Method of weighing feed water, 15.
Mixture of Anthracite and bituminous coal, Tests with, 75, 79, 81, 84, 91, 101, 109, 115, 118, 123, 142, 150, 166, 194, 197.

Object of boiler tests in general, 11.

Pea coal, Tests with Anthracite, 110, 118, 178, 219.
Petroleum oil from Canada, Test with, 151.
Petroleum, Tests with, 50, 97, 150.
Pipes beneath boiler shell in furnace, Effect of, upon economy, 67, 142.
Plain cylinder boilers, 39, 174, 177, 179.
Plant of two cast-iron sectional boilers, 215.
Plant of ten cast-iron sectional boilers with flue heater, 213.
Plant of three 60-inch double-deck boilers, 162.
Plant of four 54-inch double-deck boilers, 168.
Plant of four 48-inch double-deck boilers with flue heater, 171.
Plant of six Galloway boilers, 180.
Plant of three 48-inch horizontal return tubular boilers, 116.
Plant of six 48-inch horizontal return tubular boilers, 73.
Plant of two 60-inch horizontal return tubular boilers, 126.
Plant of three 60-inch horizontal return tubular boilers, 146.
Plant of four 60-inch horizontal return tubular boilers, 122.
Plant of six 60-inch horizontal return tubular boilers, 84.
Plant of three 66-inch horizontal return tubular boilers, 138.
Plant of two 72-inch horizontal return tubular boilers, 132.
Plant of three 72-inch horizontal return tubular boilers, 127.
Plant of two 48-inch horizontal return tubular boilers and one furnace, 70.

INDEX.

279

Plant of six 60-inch horizontal return tubular boilers with double passage for products of combustion, 75.
Plant of two 54-inch horizontal return tubular boilers with flue heater, 140.
Plant of two 60-inch horizontal return tubular boilers with flue heater, 142.
Plant of four plain cylinder boilers, 174.
Plant of five vertical tubular boilers, 196.
Plant of two vertical tubular fire box boilers with large heating surface, 208.
Plant of two vertical tubular fire-box boilers, without superheating, 203.
Plant of two vertical tubular nest boilers, 199.
Plant of two vertical tubular rolling-pin boilers with flue heater, 205.
Plant of two water-tube boilers, 231.
Plant of two water-tube boilers with flue heater, 224.
Plant of four water-tube boilers with coil of pipe in flue, 219.
Power, Relative, with different fuels, 48, 80, 91, 102.
Prevention of smoke, 53, 62, 113, 122, 137, 159.
Prevention of smoke by admitting air above fuel, 53, 62, 113, 137, 159.
Proportion of heating surface to grate surface, Effect of, upon economy, 32.
Proportion of heating surface for best results, 32.
Proportion of tube opening to grate surface, Effect of, upon economy, 34.
Proportions of air space in grates, Tests with different, 60, 95.

Quality of steam, Method of determining, 17.
Quality of steam, Calorimeter tests of the, 62.

Rapid combustion, Economy of, 183, 187, 198.
Rate of combustion, Effect of changing, 103, 106, 108, 168, 176, 189, 198, 202.
Relative amount of power developed with different fuels, 48, 80, 91, 102.
Relative economy of different kinds of boilers, 36,
Relative economy of different kinds of fuel, 43, 80, 101, 109, 115, 118, 150, 163, 166, 194, 197.
Relative labor of firing different coals, 80, 119, 196.

Screenings burned with forced draught, Tests with, 86, 190.
Sectional boilers, Cast iron, 211, 213, 215.
Shape of bridge wall, Effect of, upon economy, 51.
Size of shell, Effect of, upon economy, 34.
Size of tubes, Effect of, 35, 134.
Smokeless furnace, 136.
Steam heating surface at top of shell, Effect of, 67, 93.
Stove coal, Tests with, 115.
Summary of Tests, 234.
Summary of Tests with flue heaters, 57.
Superheated steam, Boilers producing, 67, 154, 183, 187, 190, 192, 196, 199, 201, 205, 209, 211, 215.
Superheater with horizontal boiler and auxiliary furnace, 67.
Superheating, Allowance for, 22, 186.

Superheating boilers, Comparison between saturated steam boilers and, 22.
Superheating calorimeter, 18.
Superheating steam in an independent superheater, 28, 69.
Superheating to different degrees with cast iron sectional boiler, 211.
Superheating with cast iron sectional boiler, 215.
Superheating with horizontal tubular boiler, 154.

Table No. 1, Results of tests of horizontal tubular boilers, 24.
Table No. 2, Results of tests of vertical tubular boilers, 25.
Table No. 3, Results of tests of horizontal tubular boilers with high flue temperature, anthracite coal, 29.
Table No. 4, Results of tests of horizontal tubulars boilers, with low flue temperature, anthracite coal, 29.
Table No. 5, Results of tests of horizontal tubular boilers with Cumberland coal, 30.
Table No. 6, Results of tests of water tube boilers, 41.
Table No. 7, Comparative results of tests with different fuels, referred to anthracite coal, 44.
Table No. 8, Comparative cost of coal and labor with different fuels, for 1000 H. P. plant, 47.
Table No. 9, Summary of tests with flue heaters, 57.
Table No. 10, Influence of flue heater on the draught, 59.
Temperature and draught suction, Method of determining, 16.
Tube opening to grate surface, Effect of proportion of, upon economy, 34.
Tubes, Effect of size of, upon economy, 35.

Use of additional heating surface applied to boiler in furnace, 61.

Value of superheated steam for motive power, 23.
Vertical tubular boilers, 22, 39, 183, 187, 190, 192, 196, 199, 201, 203, 205, 208.
Vertical tubular boiler with interior chamber, 187, 190.
Vertical tubular boilers with flue heater, 205.
Vertical tubular fire-box boilers, 27, 187, 203, 208.
Vertical tubular rolling-pin boilers, 183, 192, 201.

Water leg front, Boilers with, 38, 89, 107.
Water tube boilers, 217, 219, 221, 222, 224, 228, 230, 231.
Water tube boilers, Relative economy of, 40.
Wet steam produced by discharging water into steam space, 72.
Wetting bituminous coal, Economy of, 62, 167.

www.ingramcontent.com/pod-product-compliance
Lightning Source LLC
Chambersburg PA
CBHW032114230426
43672CB00009B/1731